Dissipative, Entropy-Production Systems across Condensed Matter and Interdisciplinary Classical vs. Quantum Physics

Dissipative, Entropy-Production Systems across Condensed Matter and Interdisciplinary Classical vs. Quantum Physics

Editor

Adam Gadomski

MDPI • Basel • Beijing • Wuhan • Barcelona • Belgrade • Manchester • Tokyo • Cluj • Tianjin

Editor
Adam Gadomski
Bydgoszcz University of Science and Technology
Poland

Editorial Office
MDPI
St. Alban-Anlage 66
4052 Basel, Switzerland

This is a reprint of articles from the Special Issue published online in the open access journal *Entropy* (ISSN 1099-4300) (available at: https://www.mdpi.com/journal/entropy/special_issues/Polish_Physicists_Congress_2021).

For citation purposes, cite each article independently as indicated on the article page online and as indicated below:

LastName, A.A.; LastName, B.B.; LastName, C.C. Article Title. *Journal Name* **Year**, *Volume Number*, Page Range.

ISBN 978-3-0365-5275-0 (Hbk)
ISBN 978-3-0365-5276-7 (PDF)

Cover image courtesy of Adam Gadomski

Contents

About the Editor

Adam Gadomski

Adam Gadomski, a full professor of physics at the Bydgoszcz University of Science and Technology in Poland and head of the Institute of Mathematics and Physics at the university, leads a group of academics focused on the modeling of physicochemical and biophysical processes.

He specializes in statistical soft-condensed matter physics and works in computational physics and physical computation problems, with an emphasis placed on biomaterials with a structure–property and function relationship.

Preface to "Dissipative, Entropy-Production Systems across Condensed Matter and Interdisciplinary Classical vs. Quantum Physics"

The XLVII Congress of Polish Physicists was, for the first time in over hundred years in the history of the Polish Physical Society, held in Bydgoszcz, Poland, on 19–23 September 2021 (http://47zfp.utp.edu.pl/en/47th-congress-of-the-polish-physical-society/ (accessed on 26 August 2022)), gathering in total more than two hundred and fifty regular participants and more than forty invited speakers. Amongst them, one may find the name of a 2004 Nobel Prize winner, Professor Frank Wilczek, with his partly Polish roots and special online contribution titled "Time Crystals". Over the course of five days and around fifteen sessions, covering topics ranging from the teaching and popularization of physics, the history of physics, and general physics to the physics of high energy and cosmology, gravitation, and astrophysics, the participants presented their recent contributions to the field. One of those sessions was titled "Statistical, Nonlinear and Complex Systems' Physics", and another one, relevant for this book, was the session titled "Condensed Matter Physics". (Two other sessions on quantum information and photonics, as well as on biological and interdisciplinary physics, complement the survey of subdisciplines covered.)

The book has collected ten papers (plus an editorial) addressing a range of topics in condensed matter and interdisciplinary classical vs. quantum physics. All of them are linked together by statistical physics/mechanics methods. The selection emerged from a survey of topics presented at the XLVII Congress of Polish Physicists held in Bydgoszcz. The topics address problems of classical and quantum statistical physics. As for the classical side, these comprise: multistability and ergodicity breaking; the formation of dissipative (e.g., spherulitic) structures in a nonequilibrium bath; ion-influenced conformation of biopolymers (glycosaminoglycans) near biosurfaces; infectious-disease propagation in terms of social distancing; far-from-equilibrium information processing employing networks of chemical oscillators; and a problem of equivalence of the Carnot principle to the second law of thermodynamics and non-equivalence to the first Clausius and Kelvin principles. As for the quantum side, in turn, these include: mixedness, coherence and entanglement in three-qubit states; (anti)coherence caused by a thermal bath; and, finally, quantum graphs split at vertices as simulated experimentally by using microwave networks. All contributions can be united under the common flagship of maximum-entropy and entropy production principles experienced by the respective classical and quantum systems in (non)equilibrium conditions.

The book has focused on relevant and fundamental issues of statistical classical/quantum physics (and related subdisciplines), pointing to maximum-entropy and entropy production and/or the spread of information principles experienced by the respective classical and quantum systems in (non)equilibrium conditions. The studies presented in ten chapters, also including the editorial, disclose both the theoretical depth as well as the practical usefulness of the applied classical and quantum approaches.

Adam Gadomski
Editor

entropy

Editorial

Dissipative, Entropy Production Systems across Condensed Matter and Interdisciplinary Classical vs. Quantum Physics

Adam Gadomski

Faculty of Chemical Technology and Engineering, Institute of Mathematics and Physics, Group of Modeling of Physicochemical Processes, Bydgoszcz University of Science and Technology, 85-796 Bydgoszcz, Poland; agad@pbs.edu.pl

This Special Issue collected ten papers addressing a range of topics in condensed matter and interdisciplinary classical vs. quantum physics. All of them are linked together by statistical physics/mechanics methods. The selection emerged from a survey of topics presented at the XLVII Congress of Polish Physicists held in Bydgoszcz, Poland on the 19–23 September 2021 (http://47zfp.utp.edu.pl/en/47th-congress-of-the-polish-physical-society/ (accessed on 4 August 2022)). The topics address problems of classical (CL) and quantum (QU) statistical physics. As for the CL side, these comprise: multistability and ergodicity breaking; the formation of dissipative (e.g., spherulitic) structures in a nonequilibrium bath; ion-influenced conformation of biopolymers (glycosaminoglycans) near biosurfaces; infectious-disease propagation in terms of social distancing; far-from-equilibrium information processing employing networks of chemical oscillators; and a problem of equivalence of the Carnot principle to the second law of thermodynamics, and non-equivalence to the first Clausius and Kelvin principles. As for the QU side, in turn, these include: mixedness, coherence and entanglement in three-qubit states; (anti)coherence caused by a thermal bath; and finally, quantum graphs split at vertices as simulated experimentally by using microwave networks. All contributions can be united under the common flagship of maximum-entropy and entropy production principles experienced by the respective CL and QU systems in (non)equilibrium conditions.

The XLVII Congress of Polish Physicists was, for the first time in over hundred years in the history of the Polish Physical Society, held in Bydgoszcz, Poland on the 19–23 September 2021 (http://47zfp.utp.edu.pl/en/47th-congress-of-the-polish-physical-society/ (accessed on 4 August 2022)) as a hybrid event, gathering in total more than two hundred fifty regular participants and more than forty invited speakers. Amongst them, one may find the name of a Nobel Prize winner (2004), Professor Frank Wilczek, with his special online contribution titled "Time Crystals". Over the course of five days and around fifteen sessions, covering topics ranging from physics' teaching and popularization, the history of physics, and general physics to the physics of high energy and cosmology, gravitation, and astrophysics, the participants presented their recent contributions to the field. One of those sessions was titled Statistical, Nonlinear and Complex Systems' Physics, whereas the other one, relevant for this Special Issue (SI), was the session titled Condensed Matter Physics. (Two other sessions on quantum information and photonics, as well as on biological and interdisciplinary physics, complete the selection of valuable contributions to this SI.) However, from these two sessions, one may expect to have a topical reflection contained in the overarching SI.

The thematic range of the selected contributions seems to be broad and can loosely be described as polydispersive. For example, it resembles a polynuclear path of yielding (poly)crystals. It can be so taken when looking at it from the first side. However, a closer inspection of the SI material gives rise to a much more monodispersive/single-crystal and compacted (than crudely expected) picture of the SI contents presented to a potential reader. Namely, all contributions collected can be united under the common denominator

Citation: Gadomski, A. Dissipative, Entropy Production Systems across Condensed Matter and Interdisciplinary Classical vs. Quantum Physics. *Entropy* **2022**, *24*, 1094. https://doi.org/10.3390/e24081094

Received: 4 August 2022
Accepted: 8 August 2022
Published: 9 August 2022

of maximum-entropy and entropy production principles experienced by both classical (CL) and quantum (QU) systems in (non)equilibrium conditions. In what follows, let us unveil the main messages conveyed by the ten collected papers of the SI. The proposed order of presenting the material commences with properly subordinated CL systems (seven contributions) and ends up with three remaining QU systems, presented by their authors.

The paper "Velocity Multistability vs. Ergodicity Breaking in a Biased Periodic Potential" [1] investigates the concept of multistability, i.e., the coexistence of several attractors for a given set of system parameters—it is one of the most important phenomena occurring in dynamical systems. The authors consider it in the velocity dynamics of a Brownian particle, driven by thermal fluctuations and moving in biased periodic-potential conditions. At non-zero temperatures, the ergodicity is roughly restored, whereas in low-temperature conditions, as occur in chaotic systems, the velocity dynamics are continuously affected by initial conditions due to weak ergodicity breaking. For moderate and high temperatures, the multistability is preserved with respect to the choice of both the starting position and velocity of the Brownian particle.

In the paper titled "Accumulation of Particles and Formation of a Dissipative Structure in a Nonequilibrium Bath" [2] it is demonstrated that in the non-Gaussian (Lévy noise) conditions, an inhomogeneous distribution of particles can be seen as a dissipative structure, i.e., a lower-entropy steady state of a system that permits the throughput of energy and the concurrent production of entropy. After the mechanism that maintains nonequilibrium is switched off, the relaxation stepping back to homogeneity gives rise to an increase in entropy and a corresponding decrease in free energy. Intuitive analytical formulae are proposed to connect the nonequilibrium Lévy noise and active particle behavior to the entropy decrease and energy increase.

In the study "Spherulites: How Do They Emerge at an Onset of Nonequilibrium Kinetic-Thermodynamic and Structural Singularity Addressing Conditions?" [3], the authors discuss, within a non-dimensional physical–kinetic model for far-from-equilibrium growth, the formation of special polycrystalline objects, named spherulites, exhibiting different growing-mode conditions. It appears that according to the modeling offered, it is feasible to foresee that the spherulites' emergence is possible prior to a pure diffusion-controlled (poly)crystal growth. It was demonstrated that the spherulitic onset addressing factors of the two different evolution types of spherulitic growth modes, namely, diffusion-controlled growth and mass-convection-controlled growth, emerge from the modeling performed. This resulted in accepting a conclusion that a type of liaison of amorphous and crystalline phases makes the system far better compromised to the actual (sometimes, concurrent) thermodynamic–kinetic conditions.

The proposal of a prerequisite of a (paradigmatic) chemical computer is discussed in "Information Processing Using Networks of Chemical Oscillators" [4]. It is argued that a far-from-equilibrium and oscillatory chemical-reaction system can mimic a small (three "chemical qubits") network, reminiscent of a certain one-task (chemical) computing device. As an example, the author considers a small network of three coupled chemical oscillators designated to differentiate the white from the red points of the Japan flag. The results are based on computer simulations with the two-variable Oregonator model of the oscillatory Belousov–Zhabotinsky chemical reaction. An optimized network of three interacting oscillators can discriminate between the (red vs. white) colors of randomly selected points with more than 98% accuracy.

Another paper collected by the SI which refers to CL systems, "Effect of Ion and Binding Site on the Conformation of Chosen Glycosaminoglycans at the Albumin Surface" [5], tackles the problem of ion-involving conformation of biopolymers called glycosaminoglycans near certain biopolymeric surfaces. Because of its negative surface charge, albumin plays an essential role in many physiological processes, including the ability to form molecular complexes. In turn, glycosaminoglycans such as hyaluronan and/or chondroitin sulfate are key ingredients of synovial fluid involved in the boundary lubrication regime of an articular cartilage. This study [5] uncovers an impact of relatively small ions (Na^+,

Mg^{2+} and Ca^{2+}), on human serum albumin–hyaluronan/chondroitin-sulfate interactions examined by means of molecular docking followed by molecular dynamics simulations. A certain type of glycosaminoglycan binding is analyzed by using a conformational entropy approach, and several protein–polymer complexes are studied to inspect, specifically, how the binding site and presence of ions influence their affinity conditions.

The next CL-type paper [6] of the overarching SI, titled "Does Social Distancing Matter for Infectious Disease Propagation? An SEIR Model and Gompertz Law Based Cellular Automaton", introduces a stochastic synchronous cellular automaton defined on a square lattice. Its goal is the examination of a population dynamics phenomenon based on certain automaton (Gompertz's mortality law)-involving rules. The automaton rules are based on the SEIR (susceptible $\rightarrow$ exposed $\rightarrow$ infected $\rightarrow$ recovered) model with probabilistic parameters gathered from real-world data on human mortality and the characteristics of the SARS-CoV-2 disease. With computer simulations, it is shown that there is an appreciable influence of the radius of the neighborhood on the number of infected and deceased individuals in the artificial population. Symptomatically, for a wide range of interactions of exposed individuals (having neither certain hallmarks of the disease nor having been diagnosed by suitable tests), even isolation of infected agents cannot prevent the progress of disease propagation. This supports offensive testing against the disease as one of the meaningful strategies to prevent high peaks of infection in the spread of SARS-CoV-2-like diseases and those similar.

Another paper from the CL collection, "Proof of Equivalence of Carnot Principle to II Law of Thermodynamics and Non-Equivalence to Clausius I and Kelvin Principles" [7], covers a fundamental issue of the laws of thermodynamics. Based mainly on logical rules, the author considers a principal assertion that the Kelvin principle is a weaker statement than the so-called first Clausius principle, and the latter is a weaker statement than the Carnot principle, which is equivalent to the (so-named) second Clausius principle. As a result, the Kelvin principle and the first Clausius principle are not exhaustive formulations of the second law of thermodynamics. It turns out that the Carnot principle eventually becomes such a formulation. The author indicates where the methodological errors were made in the presented proof of the equivalence of the Kelvin principle and both Clausius' principles mentioned.

The paper titled "Mixedness, Coherence and Entanglement in a Family of Three-Qubit States" [8] regards the problem of a family of states representing three-qubit systems. Formulae are derived, showing the relations between linear entropy and measures of coherence such as degree of coherence by first- and second-order correlation functions. It is demonstrated that qubit–qubit states are strongly entangled when linear entropy approaches some range of values. For such states, the authors derived the conditions determining boundary values of the corresponding entropy as being well-parametrized by the measures of coherence.

The second from the small series of QU papers, "Coherence and Anticoherence Induced by Thermal Fields" [9], addresses coherence and correlations emerging between superpositions of two bosonic modes when the modes are coupled to a third intermediate mode, and are also coupled to external modes, i.e., thermal states of unequal mean photon numbers. It turns out that one of the linear superpositions of the modes, which is efficiently decoupled from the other modes, can express a perfect coherence with the other orthogonal superposition of the modes, and can at the same time exhibit anticoherence with the intermediate mode, giving rise to entanglement between the modes. Such a system can be employed to cool modes to low-temperature values. It is demonstrated that the entanglement between the directly coupled superposition and the intermediate modes may appear in a less restricted range of the number of the thermal photons, such that the modes could be tightly entangled, even at large numbers of the thermal photons.

The last in-presentation-order (QU) paper of the SI "The Generalized Euler Characteristics of the Graphs Split at Vertices" [10] argues that there is a relationship between the generalized Euler characteristic of an original graph that was split at vertices into two disconnected subgraphs, and their generalized Euler characteristics. The theoretical results are experimentally justified by employing microwave networks that mimic quantum graphs. It is shown that the evaluation of the generalized Euler characteristics allows determining the number of vertices for which case the two subgraphs were initially connected.

To summarize, this SI has focused on relevant and fundamental issues of statistical classical/quantum physics (and related subdisciplines), pointing to maximum-entropy and entropy production (and/or the spread of information) principles experienced by the respective CL and QU systems in (non)equilibrium conditions. The studies [1–10] disclose both the theoretical depth as well as the practical usefulness of the applied CL and QU approaches.

Funding: This work is supported by an internal source BN-WTiICh-11/2022 of the Bydgoszcz University of Science and Technology.

Acknowledgments: The XLVII Congress of Polish Physicists has been sponsored in part by Kujawsko-Pomorskie Voivodeship in Toruń, Poland, as well as by Bydgoszcz city. An anonymous main sponsorship helped a great deal to make the organization of the Congress effective. With deep sadness, we wish to announce that one of invited speakers of the Congress, Marek Cieplak (P.A.S. Warsaw; biological physics), who solely announced in private his contribution to the SI, passed peacefully away in December 2021.

Conflicts of Interest: The author declares no conflict of interest.

References

1. Spiechowicz, J.; Hänggi, P.; Łuczka, J. Velocity Multistability vs. Ergodicity Breaking in a Biased Periodic Potential. *Entropy* **2022**, *24*, 98. [CrossRef] [PubMed]
2. Bier, M.; Yuvan, S. Accumulation of Particles and Formation of a Dissipative Structure in a Nonequilibrium Bath. *Entropy* **2022**, *24*, 189. [CrossRef] [PubMed]
3. Siódmiak, J.; Gadomski, A. Spherulites: How Do They Emerge at an Onset of Nonequilibrium Kinetic-Thermodynamic and Structural Singularity Addressing Conditions? *Entropy* **2022**, *24*, 663. [CrossRef] [PubMed]
4. Gorecki, J. Information Processing Using Networks of Chemical Oscillators. *Entropy* **2022**, *24*, 1054. [CrossRef]
5. Sionkowski, P.; Bełdowski, P.; Kruszewska, N.; Weber, P.; Marciniak, B.; Domino, K. Effect of Ion and Binding Site on the Conformation of Chosen Glycosaminoglycans at the Albumin Surface. *Entropy* **2022**, *24*, 811. [CrossRef] [PubMed]
6. Biernacki, S.; Malarz, K. Does Social Distancing Matter for Infectious Disease Propagation? An SEIR Model and Gompertz Law Based Cellular Automaton. *Entropy* **2022**, *24*, 832. [CrossRef] [PubMed]
7. Koczan, G.M. Proof of Equivalence of Carnot Principle to II Law of Thermodynamics and Non-Equivalence to Clausius I and Kelvin Principles. *Entropy* **2022**, *24*, 392. [CrossRef] [PubMed]
8. Kalaga, J.J.K.; Leoński, W.; Szczęśniak, R.; Peřina, J., Jr. Mixedness, Coherence and Entanglement in a Family of Three-Qubit States. *Entropy* **2022**, *24*, 324. [CrossRef] [PubMed]
9. Sun, L.; Liu, Y.; Li, C.; Zhang, K.; Yang, W.; Ficek, Z. Coherence and Anticoherence Induced by Thermal Fields. *Entropy* **2022**, *24*, 692. [CrossRef] [PubMed]
10. Farooq, O.; Ławniczak, M.; Akhshani, A.; Bauch, S.; Sirko, L. The Generalized Euler Characteristics of the Graphs Split at Vertices. *Entropy* **2022**, *24*, 387. [CrossRef] [PubMed]

entropy

MDPI

Article

Velocity Multistability vs. Ergodicity Breaking in a Biased Periodic Potential

Jakub Spiechowicz [1,*], Peter Hänggi [2] and Jerzy Łuczka [1]

[1] Institute of Physics, University of Silesia in Katowice, 41-500 Chorzów, Poland; jerzy.luczka@us.edu.pl
[2] Institute of Physics, University of Augsburg, Universitätstr. 1, 86135 Augsburg, Germany; hanggi@physik.uni-augsburg.de
* Correspondence: jakub.spiechowicz@us.edu.pl

Abstract: Multistability, i.e., the coexistence of several attractors for a given set of system parameters, is one of the most important phenomena occurring in dynamical systems. We consider it in the velocity dynamics of a Brownian particle, driven by thermal fluctuations and moving in a biased periodic potential. In doing so, we focus on the impact of ergodicity—A concept which lies at the core of statistical mechanics. The latter implies that a single trajectory of the system is representative for the whole ensemble and, as a consequence, the initial conditions of the dynamics are fully forgotten. The ergodicity of the deterministic counterpart is strongly broken, and we discuss how the velocity multistability depends on the starting position and velocity of the particle. While for non-zero temperatures the ergodicity is, in principle, restored, in the low temperature regime the velocity dynamics is still affected by initial conditions due to weak ergodicity breaking. For moderate and high temperatures, the multistability is robust with respect to the choice of the starting position and velocity of the particle.

Keywords: multistability; ergodicity; Brownian motion; tilted periodic potential

Citation: Spiechowicz, J.; Hänggi, P.; Łuczka, J. Velocity Multistability vs. Ergodicity Breaking in a Biased Periodic Potential. *Entropy* **2022**, *24*, 98. https://doi.org/10.3390/e24010098

Academic Editor: Antonio Maria Scarfone

Received: 16 December 2021
Accepted: 5 January 2022
Published: 7 January 2022

1. Introduction

Research in nonequilibrium statistical physics provides a wealth of intriguing dynamics in which phenomena that are forbidden in equilibrium states may emerge. Prominent examples include anomalous diffusion [1–4], Brownian yet non-Gaussian diffusion [5–9], noise-assisted transport [10,11], and negative mobility [12–15], to name only a few. While the behaviour of low dimensional systems, where usually only one or two attractors rule the dynamics, has been studied intensively, much less is known for systems where several attractors coexist for a given set of the system parameters. This feature, called multistability is commonly found in different areas of science such as physics, chemistry, biology, economy, and in nature [16].

In this paper, we reinvestigate in this context the paradigmatic model of nonequilibrium statistical physics, namely, underdamped Brownian motion in a biased periodic potential. This nonlinear system enjoys never ending interest as its different aspects have already been studied for several decades [17–32]. The latter are mostly focused on the diffusive properties of the system. For instance, it may exhibit unusual phenomena such as the giant diffusion [19,20,25,32] or the non-monotonic temperature dependence of a diffusion coefficient [25,30,32,33]. Both these effects are related to a bistability observed in the velocity dynamics of the system. The later effect is well known due to the work by Risken et al. [34] who found that at low friction and appropriate bias values the velocity can be stable in a locked solution (the particle is trapped in a potential minimum) but also in a running solution (the motion is unbounded in space).

Here, we focus on multistability of the Brownian velocity dynamics in a tilted periodic potential. Despite so many years of intensive research on various aspects of this setup, the latter peculiar effect has been addressed only very recently [30] and later it was explained by recoursing to the arcsine law [35], which is a cornerstone of extreme-value statistics. Specifically, we investigate the role of ergodicity breaking and its consequences on the velocity multistability. Ergodicity lies at the basis of statistical mechanics and implies that, over long enough observation times, the time averages of observables correspond to the equilibrium ensemble averages [1,36,37]. Equivalently, it states that a single trajectory is representative for the ensemble. An increasing number of systems exhibit nonergodic properties [1,36,37], in particular due to the ultra slow dynamics and non-exponential relaxation.

The paper is organized as follows. In Section 2, we recall the formulation of the model and introduce the dimensionless quantities. In Section 3, we discuss the results, in particular the effect of ergodicity breaking on the velocity multistability occurring in this paradigmatic system. Finally, Section 4 provides a discussion and concluding remarks.

2. Methods

In this work, we study dynamics of a classical inertial Brownian particle of mass M moving in a spatially periodic and symmetric potential $U(x) = U(x + L)$ of period L and subjected to a static bias F. This system can be described by the following Langevin equation

$$M\ddot{x} + \Gamma\dot{x} = -U'(x) + F + \sqrt{2\Gamma k_B T}\,\xi(t), \tag{1}$$

where the dot and prime denote differentiation with respect to the time t and the particle coordinate x, respectively. The parameter Γ is the friction coefficient, T is temperature, and k_B denotes the Boltzmann constant. We consider the potential $U(x)$ in the form

$$U(x) = -\Delta U \sin\left(\frac{2\pi}{L}x\right), \tag{2}$$

where ΔU denotes half of the potential barrier height. Thermal equilibrium fluctuations are modeled by the δ-correlated Gaussian white noise whose statistical characteristics read

$$\langle \xi(t) \rangle = 0, \quad \langle \xi(t)\xi(s) \rangle = \delta(t-s). \tag{3}$$

The noise prefactor $2\Gamma k_B T$ satisfies the fluctuation–dissipation theorem that ensures the canonical Gibbs statistics when the system is at the equilibrium state.

The above Langevin Equation (1) can be transformed into the dimensionless form

$$\ddot{\hat{x}} + \gamma\dot{\hat{x}} = -\mathcal{U}'(\hat{x}) + \sqrt{2\gamma\theta}\,\hat{\xi}(\hat{t}) \tag{4}$$

by introducing the rescaled coordinate $\hat{x}$ and time $\hat{t}$,

$$\hat{x} = \frac{2\pi}{L}x, \quad \hat{t} = \frac{t}{\tau_0}, \quad \tau_0 = \frac{L}{2\pi}\sqrt{\frac{M}{\Delta U}}, \tag{5}$$

where the characteristic time τ_0 is proportional to the inverse of frequency ω_0 of small oscillations in the potential well of $U(x)$. The effective dimensionless potential is

$$\mathcal{U}(\hat{x}) = -\sin\hat{x} - f\hat{x}. \tag{6}$$

The dimensionless friction coefficient γ and bias f read

$$\gamma = \frac{1}{2\pi}\frac{L}{\sqrt{M\Delta U}}\Gamma, \quad f = \frac{1}{2\pi}\frac{L}{\Delta U}F. \tag{7}$$

The rescaled temperature θ is the ratio of thermal energy $k_B T$ to half of the barrier height the particle needs to overcome the original potential well, namely,

$$\theta = \frac{k_B T}{\Delta U}. \tag{8}$$

The dimensionless thermal noise $\hat{\xi}(\hat{t})$ is statistically equivalent to $\xi(t)$, meaning that it is a stationary Gaussian stochastic process with vanishing mean. Later, we use only the rescaled quantities, and therefore in order to improve the readability of notation from now on we omit the hat appearing in Equation (4).

The model of a Brownian particle moving in a washboard potential formulated in terms of the Langevin Equation (4) served for decades as a tool for the investigation of transport effects occurring in both classical and quantum systems. For instance, it has been employed for understanding the dynamics of phase across the Josephson junction [38], rotating dipoles in external fields [39], superionic conductors [40], charge density waves [41], and cold atoms dwelling in optical lattices [42–44]. Further systems are mentioned in Ref. [17].

The analytical methods of solution for the Fokker–Planck equation corresponding to Equation (4) are not yet elaborated, therefore in doing so, we rely solely on precise numerical simulations. All calculations have been completed using a Compute Unified Device Architecture (CUDA) environment implemented on a modern desktop graphics processing unit (GPU). This method allowed to speed up necessary calculations by a factor of the order 10^3 as compared to the traditional methods. We refer interested reader to Refs. [45,46] where more details on this scheme can be found. Here, we only mention that all quantities of interest were averaged over the ensemble of 2^{19} = 524,288 system trajectories.

3. Results

In this paper, we investigate various aspects of multistability in the velocity dynamics of a Brownian particle dwelling in a tilted periodic potential. This interesting phenomenon has been addressed for this setup only very recently [35], however, it has been also reported in systems driven by other types of noise. Examples include fractional Gaussian noise [29], Ornstein–Uhlenbeck, and harmonic Levy noise [47,48], to name but a few.

In Figure 1 we exemplify the velocity multistability phenomenon occurring in this system. The probability distribution $p(v)$ for the instantaneous long time velocity v obtained from the histogram of the latter quantity is depicted for fixed time $t = 10^4$ (but is time-invariant) and for different dimensionless temperatures θ. The issue of measurement of the instantaneous velocity of a Brownian particle is presented in Ref. [49]. In Figure 1, one can observe three well pronounced maxima. One of them corresponds to the velocity $v = 0$ (the locked state) and the other two with $v \neq 0$ are related to running solutions. This means that these values occur significantly more frequently than the others and therefore are more stable. This observation matches the common definition of multistability for stochastic systems [50]. We note that, as temperature θ increases, the difference between each maximum becomes less pronounced and eventually disappears.

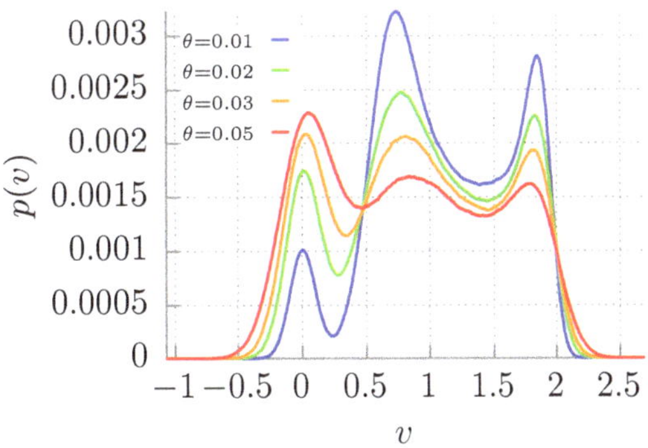

Figure 1. The probability distribution $p(v)$ for the instantaneous long time velocity $v = v(t)$ of the Brownian particle is illustrated for $t = 10^4$ and selected values of temperatures θ of the system. The used parameters read $\gamma = 0.66$ and $f = 0.91$.

In Ref. [35], the origin of the multistability effect is explained in terms of the arcsine law for the velocity dynamics at the zero temperature limit $\theta = 0$, i.e., as the trace of deterministic dynamics perturbed by thermal noise. In such a case in the long time regime, the velocity $v(t)$ of the particle is a time-periodic function. Moreover, the ergodicity of the setup is strongly broken, which means that its phase space can be divided into two non-intersecting invariant sets corresponding to the locked and running state [51]. We visualize this in Figure 2, where the time averaged particle velocity

$$\mathbf{v} = \lim_{t \to \infty} \frac{1}{t} \int_0^t ds\, \dot{x}(s) \tag{9}$$

is depicted as a function of the initial conditions for the coordinate $x(0) = x_0$ and velocity $v(0) = v_0$. The black region corresponds to the locked state with $\mathbf{v} = 0$ whereas the grey one indicates the regime of a running solution for which $\mathbf{v} \neq 0$. Therefore, different initial conditions $\{x_0, v_0\}$ can lead to a distinct average velocity $\mathbf{v}$. It is a disturbing situation, as typically in experiments the initial conditions are not known a priori or can be settled only with a finite resolution. To get rid of the dependence of the obtained results on the initial conditions, one needs to average over them. In Ref. [35], the authors distributed x_0 and v_0 uniformly over the intervals $[0, 2\pi]$ and $[-2, 2]$, respectively. Moreover, they found that in such a case the initial conditions induce an almost uniformly distributed phase shift φ in the time-periodic dependence of the velocity $v(t)$ in the long time regime. This in turn results in the arcsine law for the velocity probability density $p(v)$ which constitutes the backbone of multistability in this system.

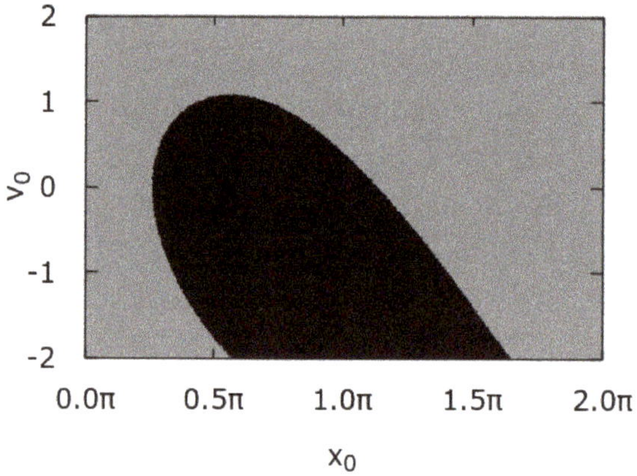

Figure 2. The basins of attraction for the time averaged velocity $\mathbf{v}$ of the particle. The black colour codes the locked state $\mathbf{v} = 0$ whereas the grey part indicates the regime with running solutions $\mathbf{v} \neq 0$. Parameters read $\gamma = 0.66$, $f = 0.91$ and $\theta = 0$.

In this work, we present a complementary study. Namely, we investigate in detail the influence of various distributions of initial conditions $\{x_0, v_0\}$ on the velocity multistability phenomenon. In Figure 3, we show the probability distribution $p(v)$ for the instantaneous long time velocity v of the Brownian particle for the deterministic system $\theta = 0$ and different choice of the initial conditions. In simulations, the moment of time is fixed $t_i = 10^4$. In the inset we depict the corresponding probability distribution $P(\mathbf{v})$ for the time averaged velocity $\mathbf{v}$. In panel (a) the initial position and velocity are fixed, $x_0 = 0, v_0 = 0$. The corresponding probability densities are represented by the Dirac-delta $p_{x_0}(x) = \delta(x)$ and $p_{v_0}(v) = \delta(v)$, respectively. Consequently, as the system is noiseless $\theta = 0$, the resulting probability distributions $p(v)$ and $P(\mathbf{v})$ for the instantaneous long time v and time averaged velocity $\mathbf{v}$, respectively, take the Dirac-delta forms. All phase space trajectories follow the same route, and the multistability effect is absent. The situation changes drastically already if the initial position of the particle is distributed uniformly over the period $L = 2\pi$ of the potential $U(x)$, i.e., $p_{x_0}(x) = \mathsf{U}(0, 2\pi)$ (see panel (b)). Here, $\mathsf{U}(a, b)$ indicates the uniform distribution over the interval $[a, b]$. The starting velocity of the particle can be fixed $p_{v_0}(v) = \delta(v)$ but still the systems display multimodality in the probability density $p(v)$. In fact, in such a case, even four distinct maxima are visible there. In the inset, we note that both locked $\mathbf{v} = 0$ and running $\mathbf{v} = 0$ states are represented in the ensemble of system trajectories. If we permute the initial conditions, i.e., the starting coordinate $p_{x_0}(x) = \delta(x)$ but $p_{v_0}(v) = \mathsf{U}(-2, 2)$ (see panel (c)), the multistability emerges but the locked state is not sampled at all. This situation can be modified depending on the choice of the initial coordinate as it is demonstrated in panel (d) where, in contrast, $p_{x_0}(x) = \delta(x - \pi)$.

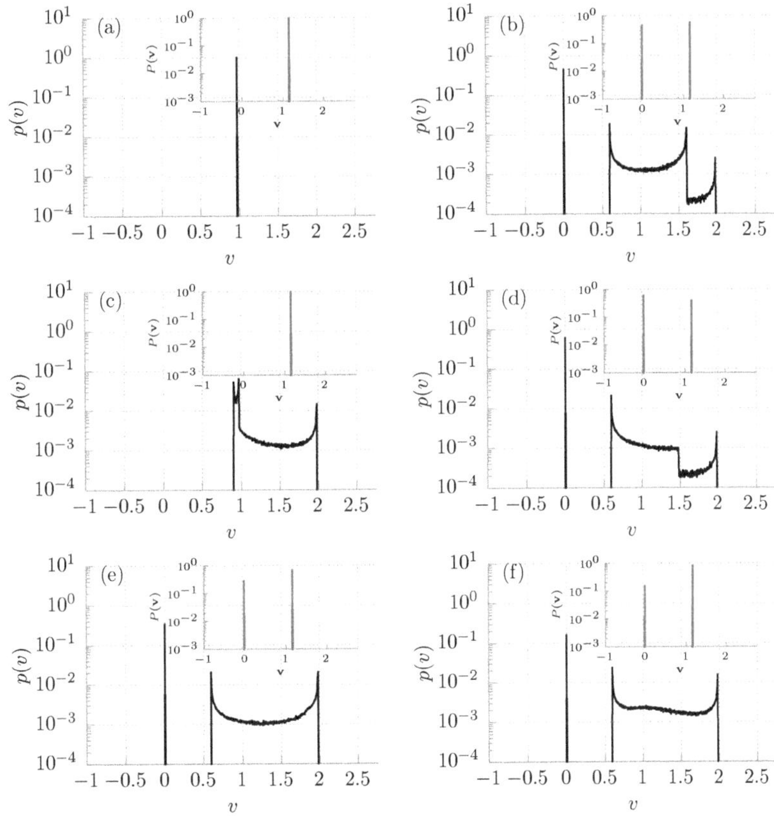

Figure 3. The probability distribution $p(v)$ for the instantaneous long time velocity v of the Brownian particle is depicted in the deterministic regime $\theta = 0$ for $t = 10^4$ and different choice of the initial conditions for the system. Panel (**a**): $p_{x_0}(x) = \delta(x)$, $p_{v_0}(v) = \delta(v)$; (**b**): $p_{x_0}(x) = \mathsf{U}(0, 2\pi)$, $p_{v_0}(v) = \delta(v)$; (**c**): $p_{x_0}(x) = \delta(x)$, $p_{v_0}(v) = \mathsf{U}(-2, 2)$; (**d**): $p_{x_0}(x) = \delta(x - \pi)$, $p_{v_0}(v) = \mathsf{U}(-2, 2)$; (**e**): $p_{x_0}(x) = \mathsf{U}(0, 2\pi)$, $p_{v_0}(v) = \mathsf{U}(-2, 2)$; and (**f**): $p_{x_0}(x) = \mathsf{N}(0, 1)$, $p_{v_0}(v) = \mathsf{N}(0, 1)$, where $\mathsf{U}(a, b)$ indicates the uniform distribution over the interval $[a, b]$. Likewise, $\mathsf{N}(\mu, \sigma^2)$ is the Gaussian distribution with the mean μ and the variance σ^2. In the inset, the corresponding probability distribution $P(\mathbf{v})$ for the time averaged velocity $\mathbf{v}$ is shown. Parameters read $\gamma = 0.66$, $f = 0.91$, and $\theta = 0$.

Overall, if the ergodicity of the system is broken, the initial conditions are never forgotten and therefore crucially impact the results. Depending on the circumstances, this behaviour may be seen as a feature, not a bug. Nevertheless, the only way to cure it is to properly average over the initial conditions. In doing so, each of them must be taken into account equally; none can be preferred. This requirement translates to the fact that initial conditions must be equally probable and therefore uniformly distributed over the whole phase space. As the system under consideration is spatially periodic $U(x) = U(x + L)$, the periodic boundary condition can be employed to yield $p_{x_0}(x) = \mathsf{U}(0, 2\pi)$. It is not the case for the starting velocity v_0 of the particle which in principle is unbounded. However, naturally such a situation cannot be implemented in numerical simulations, and therefore one needs to carefully check the impact of the initial velocity subspace volume on the obtained results. As we demonstrated, if this is not performed thoroughly one can significantly spoil the outcomes and, e.g., break the inherent symmetries of the system [52]. We checked that, in the considered regime, the condition $p_{v_0}(v) = \mathsf{U}(-2, 2)$ is sufficient and that further increase in the initially chosen velocity subspace volume would not alter the outcomes. In Figure 3e, we reproduce the result from Ref. [35] obtained for $p_{x_0}(x) = \mathsf{U}(0, 2\pi)$ and $p_{v_0}(v) = \mathsf{U}(-2, 2)$. The characteristic U-shape part which portrays the arcsine law corresponding to the running state is visible in the probability density $p(v)$. Consequently, the velocity dynamics is multistable.

One can claim that the initial conditions, especially the velocity, should be distributed according to the Gaussian probability density, as then it obeys the canonical Gibbs statistics (Maxwell–Boltzmann distribution) valid for equilibrium systems. Obviously, such a choice does not satisfy the above discussed condition of equal probability. In panel (f), we show that, as a consequence of the non-uniformity for $p_{x_0}(x) = \mathsf{N}(0, 1)$ and $p_{v_0}(v) = \mathsf{N}(0, 1)$, where $\mathsf{N}(\mu, \sigma^2)$ is the Gaussian distribution with the mean μ and the variance σ^2, the results are deformed and the arcsine law is not properly recovered. There is one more argument that the condition of equal probability is the only one to be correct and consistent with the case of non-zero temperature. In the running state, the long time velocity trajectory $v(t)$ is a periodic function of time and can be well approximated by the simple periodic function [35]

$$V(t) = A \sin(\omega t + \phi) + c. \tag{10}$$

For a fixed set of the system parameters, the constants (A, ω, c) are the same for all initial conditions $\{x_0, v_0\}$. However, the distribution of the phase shift ϕ depends on the distribution of initial conditions $\{x_0, v_0\}$. This fact is reflected in different probability densities $p(v)$ for the instantaneous velocity depicted in Figure 3. As the ergodicity of the system is broken, the distributions $p(v)$ generally depend on the measurement time $t = t_i$. The exception is the uniform distribution for the phase ϕ corresponding to the panel (e) in Figure 3 for which $p(v)$ is time-invariant [35]. The latter feature is characteristic for ergodic systems and is crucial from the experimental point of view.

As we just reported, the ergodicity of the deterministic system with $\theta = 0$ is broken for the parameter regimes in which it exhibits the multistability phenomenon [32]. One may argue that the case $\theta = 0$ is only an idealization and, in practice, there exists no realistic situation with zero temperature. However, the ergodicity breaking in a deterministic system often also carries prominent consequences for non-zero temperature. In particular, for any positive temperature $\theta > 0$ the system described by Equation (4) is always ergodic, although it is not a trivial fact, as it is driven by noise [53]. At non-zero temperatures, the whole phase space is accessible due to thermally activated escape events connecting the coexisting deterministic disjoint attractors. However, if the temperature tends to zero $\theta \to 0$, the time τ after it is fully sampled becomes extremely long and goes to infinity as $\tau \to \infty$. From an experimental point of view, due to finite observation time, the system seemingly behaves as being non-ergodic although in fact it is ergodic. Such a situation is often termed as weak ergodicity breaking [36,37,51] and can be identified with an unusually slow relaxation of the system towards its steady state which manifests itself as the nonequivalence of time and ensemble averages. In the latter case, the initial conditions do not fade, but in fact modify the results. We exemplify this feature in Figure 4a, where we depict the probability distribution $p(v)$ for the instantaneous long time velocity v of the Brownian particle for different initial conditions and low temperature $\theta = 0.0001$. Clearly, when the particle starts from $x_0 = 0$ and $v_0 = 0$ (see the red solid line), even in the long time limit, there are only running solutions. On the other hand, if the initial position x_0 and velocity v_0 are either uniformly or normal distributed (see the blue or green line, respectively), the multistability emerges, but one can still note quantitative difference between these two initial conditions. In contrast, in panel (b) we depict the same characteristics but for higher temperature $\theta = 0.05$. Then, thermal fluctuations are strong enough to recover the ergodicity of the system and there are no longer differences between different initial conditions. Even when the particle trajectories start from the same point in the phase space $x_0 = 0$, $v_0 = 0$ (see the red solid line), the whole density is obtained.

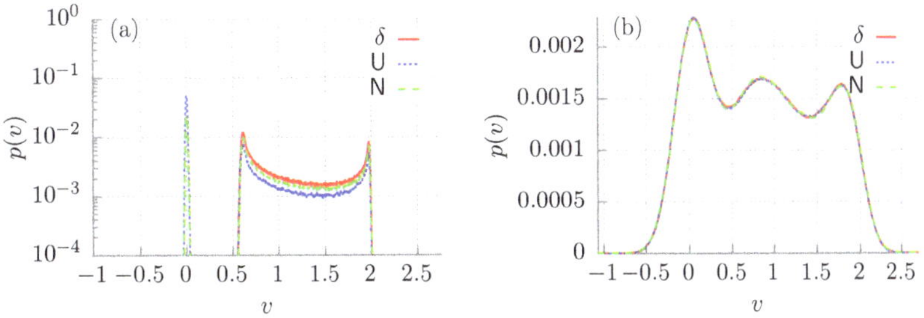

Figure 4. The probability distribution $p(v)$ for the instantaneous long time velocity v of the Brownian particle is depicted for $t = 10^4$ and different initial conditions of the system. The red solid line indicates $p_{x_0}(x) = \delta(x)$, $p_{v_0}(v) = \delta(v)$. The blue dotted line corresponds to $p_{x_0}(x) = \mathsf{U}(0, 2\pi)$, $p_{v_0}(v) = \mathsf{U}(-2, 2)$. The green dashed line denotes $p_{x_0}(x) = \mathsf{N}(0, 1)$, $p_{v_0}(v) = \mathsf{N}(0, 1)$. In panel (**a**) temperature is $\theta = 0.0001$ while in (**b**) $\theta = 0.05$. Other parameters read $\gamma = 0.66$, $f = 0.91$.

4. Discussion

In conclusion, we thoroughly investigated the influence of initial conditions distribution on the multistability of velocity dynamics for the Brownian particle in a tilted periodic potential. The ergodicity of the deterministic system is strongly broken, and therefore the initial conditions are never forgotten and crucially impact the obtained results. The only way to correctly sample the whole state space of the system is to average over them in such a way that no single one is preferred. The latter condition translates to a uniform distribution in the initial phase space of the system. We demonstrated that, while for non-zero temperatures the ergodicity is in principle restored, in a low temperature regime the results are still significantly affected by the initial conditions due to the weak ergodicity breaking. This means that the time needed for the ergodicity reinstatement tends to infinity when temperature goes down to zero. For moderate and high temperature regimes, the detected multistability is robust with respect to the choice of initial conditions. It is valid even when the whole ensemble starts from a given point in the phase space of the system. A remaining question is how the time needed for the ergodicity restoration depends on temperature. This constitutes a challengeable objective which we hope to address in the future.

Author Contributions: Conceptualization, J.S.; Investigation, J.S.; Software, J.S.; Visualization, J.S.; Writing—original draft, J.S.; Writing—review & editing, J.S., P.H. and J.Ł. All authors have read and agreed to the published version of the manuscript.

Funding: This work has been supported by the Grant NCN No. 2017/26/D/ST2/ 00543 (J.S.).

Institutional Review Board Statement: Not applicable.

Informed Consent Statement: Not applicable.

Data Availability Statement: The data that support the findings of this study are available from the corresponding author upon reasonable request.

Acknowledgments: P.H. and J.S. wish to acknowledge the financial support obtained via the University of Augsburg Guestprogramme funded by Bavarian Ministery for Science and Art, Germany.

Conflicts of Interest: The authors declare no conflict of interest.

References

1. Metzler, R.; Jeon, J.H.; Cherstvy, A.G.; Barkai, E. Anomalous diffusion models and their properties: Non-stationarity, non-ergodicity, and ageing at the centenary of single particle tracking. *Phys. Chem. Chem. Phys.* **2014**, *16*, 24128. [CrossRef]
2. Spiechowicz, J.; Łuczka, J. Subdiffusion via dynamical localization induced by thermal equilibrium fluctuations. *Sci. Rep.* **2017**, *7*, 16451. [CrossRef]
3. Spiechowicz, J.; Łuczka, J. SQUID ratchet: Statistics of transitions in dynamical localization. *Chaos* **2019**, *29*, 013105. [CrossRef] [PubMed]
4. Spiechowicz, J.; Hänggi, P.; Łuczka, J. Coexistence of absolute negative mobility and anomalous diffusion. *New J. Phys.* **2019**, *21*, 083029. [CrossRef]
5. Leptos, K.C.; Guasto, J.S.; Gollub, J.P.; Pesci, A.I.; Goldstein, R.E. Dynamics of Enhanced Tracer Diffusion in Suspensions of Swimming Eukaryotic Microorganisms. *Phys. Rev. Lett.* **2009**, *103*, 198103. [CrossRef]
6. Wang, B.; Kuo, J.; Bae, S.C.; Granick, S. When Brownian diffusion is not Gaussian. *Nat. Mater.* **2012**, *11*, 481. [CrossRef] [PubMed]
7. Chechkin, A.V.; Seno, F.; Metzler, R.; Sokolov, I.M. Brownian yet non-Gaussian diffusion: From superstatistics to subordination of diffusing diffusivities. *Phys. Rev. X* **2017**, *7*, 021002. [CrossRef]
8. Białas, K.; Łuczka, J.; Hänggi, P.; Spiechowicz, J. Colossal Brownian yet non-Gaussian diffusion induced by nonequilibrium noise. *Phys. Rev. E* **2020**, *102*, 042121. [CrossRef] [PubMed]
9. Barkai, E.; Burov, S. Packets of spreading particles exhibit universal exponential tails. *Phys. Rev. Lett.* **2020**, *124*, 060603. [CrossRef]
10. Hänggi, P.; Marchesoni, F. Artificial Brownian motors: Controlling transport on the nanoscale. *Rev. Mod. Phys.* **2009**, *81*, 387. [CrossRef]
11. Slapik, A.; Łuczka, J.; Spiechowicz, J. Temperature-induced tunable particle separation. *Phys. Rev. Appl.* **2019**, *12*, 054002. [CrossRef]
12. Machura, Ł.; Kostur, M.; Talkner, P.; Łuczka, J.; Hänggi, P. Absolute negative mobility induced by thermal equilibrium fluctuations. *Phys. Rev. Lett.* **2007**, *98*, 040601. [CrossRef] [PubMed]
13. Nagel, J.; Speer, D.; Gaber, T.; Sterck, A.; Eichhorn, R.; Reimann, P.; Ilin, K.; Siegel, M.; Koelle, D.; Kleiner, R. Observation of negative absolute resistance in a Josephson junction. *Phys. Rev. Lett.* **2008**, *100*, 217001. [CrossRef] [PubMed]
14. Slapik, A.; Łuczka, J.; Hänggi, P.; Spiechowicz, J. Tunable mass separation via negative mobility. *Phys. Rev. Lett.* **2019**, *122*, 070602. [CrossRef] [PubMed]
15. Slapik, A.; Spiechowicz, J. Tunable particle separation via deterministic absolute negative mobility. *Sci. Rep.* **2020**, *10*, 16639. [CrossRef] [PubMed]
16. Pisarchik, A.; Feudel, U. Control of multistability. *Phys. Rep.* **2014**, *540*, 167. [CrossRef]
17. Risken, H. *The Fokker-Planck Equation: Methods of Solution and Applications*; Springer: Berlin/Heidelberg, Germany, 1996.
18. Lindner, B.; Kostur, M.; Schimansky-Geier, L. Optimal diffusive transport in a tilted periodic potential. *Fluct. Noise Lett.* **2001**, *R25*, 173. [CrossRef]
19. Reimann, P.; Van den Broeck, C.; Linke, H.; Hänggi, P.; Rubi, J.M.; Perez-Madrid, A. Giant acceleration of free diffusion by use of tilted periodic potentials. *Phys. Rev. Lett.* **2001**, *87*, 010602. [CrossRef]
20. Reimann, P.; Van den Broeck, C.; Linke, H.; Hänggi, P.; Rubi, J.M.; Perez-Madrid, A. Diffusion in tilted periodic potentials: Enhancement, universality, and scaling. *Phys. Rev. E* **2001**, *65*, 031104. [CrossRef]
21. Constantini, G.; Marchesoni, F. Threshold diffusion in a tilted washboard potential. *Europhys. Lett.* **1999**, *48*, 491. [CrossRef]
22. Lindenberg, K.; Lacasta, A.M.; Sancho, J.M.; Romero, A.H. Transport and diffusion on crystalline surfaces under external forces. *New. J. Phys.* **2005**, *7*, 29. [CrossRef]
23. Marchenko, I.G.; Marchenko, I.I. Diffusion in the system with low dissipation: Exponential growth with temperature. *Europhys. Lett.* **2012**, *100*, 50005. [CrossRef]
24. Latorre, J.C.; Pavliotis, G.A.; Kramer, P.R. Corrections to Einstein's relation for Brownian motion in a tilted periodic potential. *J. Stat. Phys.* **2013**, *150*, 776. [CrossRef]
25. Lindner, B.; Sokolov, I.M. Giant diffusion of underdamped particles in a biased periodic potential. *Phys. Rev. E* **2016**, *93*, 042106. [CrossRef] [PubMed]
26. Zhang, J.M.; Bao, J.D. Transition of multidiffusive states in a biased periodic potential. *Phys. Rev. E* **2017**, *95*, 032107. [CrossRef] [PubMed]
27. Marchenko, I.G.; Marchenko, I.I.; Tkachenko, V.I. Temperature-Abnormal Diffusivity in Underdamped Spatially Periodic Systems. *JETP Lett.* **2017**, *106*, 242. [CrossRef]
28. Cheng, C.; Cirillo, M.; Salina, G.; Gronbech-Jensen, N. Nonequilibrium transient phenomena in the washboard potential. *Phys. Rev. E* **2018**, *98*, 012140. [CrossRef]
29. Goychuk, I. Fractional Hydrodynamic Memory and Superdiffusion in Tilted Washboard Potentials. *Phys. Rev. Lett.* **2019**, *123*, 180603. [CrossRef]
30. Spiechowicz, J.; Łuczka, J. Diffusion in a biased washboard potential revisited. *Phys. Rev. E* **2020**, *101*, 032123. [CrossRef]
31. Goychuk, I.; Pöschel, T. Nonequilibrium Phase Transition to Anomalous Diffusion and Transport in a Basic Model of Nonlinear Brownian Motion. *Phys. Rev. Lett.* **2021**, *127*, 110601. [CrossRef] [PubMed]

32. Spiechowicz, J.; Łuczka, J. The conundrum of weak noise limit for diffusion in a tilted periodic potential. *Phys. Rev. E* **2021**, *104*, 034104. [CrossRef] [PubMed]
33. Marchenko, I.G.; Marchenko, I.I.; Tkachenko, V.I. Temperature-Abnormal Diffusivity in Tilted Periodic Potentials. *JETP Lett.* **2019**, *109*, 671. [CrossRef]
34. Vollmer, H.; Risken, H. Eigenvalues and their connection to transition rates for the Brownian motion in an inclined cosine potential. *J. Phys. B. Cond. Mat.* **1983**, *52*, 259. [CrossRef]
35. Spiechowicz, J.; Łuczka, J. Arcsine law and multistable Brownian dynamics in a tilted periodic potential. *Phys. Rev. E* **2021**, *104*, 024132. [CrossRef] [PubMed]
36. Meroz, Y.; Sokolov, I.M. A toolbox for determining subdiffusive mechanisms. *Phys. Rep.* **2015**, *573*, 1–29. [CrossRef]
37. Bouchaud, J.P. Weak ergodicity breaking and aging in disordered systems. *J. Phys. I France* **1992**, *2*, 1705. [CrossRef]
38. Barone, A.; Paternò, G. *Physics and Application of the Josephson Effect*; Wiley: New York, NY, USA, 1982.
39. Coffey, W.T.; Kalmykov, Y.P.; Waldron, J.T. *The Langevin Equation*; World Scientific: Singapore, 2004; See Sects. 5 and 7–10 therein.
40. Grüner, G.; Zawadowski, A.; Chaikin, P.M. Nonlinear Conductivity and Noise due to Charge-Density-Wave Depinning in NbSe3. *Phys. Rev. Lett.* **1981**, *46*, 511. [CrossRef]
41. Fulde, P.; Pietronero, L.; Schneider, W.R.; Strässler, S. Problem of Brownian Motion in a Periodic Potential. *Phys. Rev. Lett.* **1975**, *35*, 1776. [CrossRef]
42. Denisov, S.; Flach, S.; Hänggi, P. Tunable transport with broken space–Time symmetries. *Phys. Rep.* **2014**, *538*, 77. [CrossRef]
43. Kindermann, F.; Dechant, A.; Hohmann, M.; Lausch, T.; Mayer, D.; Schmidt, F.; Lutz, E.; Widera, A. Nonergodic diffusion of single atoms in a periodic potential. *Nat. Phys.* **2017**, *13*, 137. [CrossRef]
44. Dechant, A.; Kindermann, F.; Widera, A.; Lutz, E. Continuous-Time Random Walk for a Particle in a Periodic Potential. *Phys. Rev. Lett.* **2019**, *123*, 070602. [CrossRef]
45. Spiechowicz, J.; Kostur, M.; Machura, Ł. GPU accelerated Monte Carlo simulation of Brownian motors dynamics with CUDA. *Comp. Phys. Commun.* **2015**, *191*, 140. [CrossRef]
46. Seibert, A.; Denisov, S.; Ponomarev, A.; Hänggi, P. Mapping the Arnold web with a GPU-supercomputer. *Chaos* **2011**, *21*, 043123. [CrossRef]
47. Li, F.; Chen, X.S.; Liu, J. Anomalous correlated Lévy flight induced by coexistence of correlation and dissipative nonlinearity. *J. Stat. Mech.* **2020**, *2020*, 083209. [CrossRef]
48. Li, F.; Zhu, P.; Bao, J.D.; Chen, X.S.; Liu, J. Nontrivial anomalous diffusions induced by the harmonic velocity Lévy noise. *J. Stat. Mech.* **2021**, *2021*, 043210. [CrossRef]
49. Kheifets, S.; Simha, A.; Melin, K.; Li, T.; Raizen, M.G. Observation of Brownian Motion in Liquids at Short Times: Instantaneous Velocity and Memory Loss. *Science* **2014**, *343*, 1493. [CrossRef] [PubMed]
50. Kraut, S.; Feudel, U.; Grebogi, C. Preference of attractors in noisy multistable systems. *Phys. Rev. E* **1999**, *59*, 5253. [CrossRef] [PubMed]
51. Spiechowicz, J.; Łuczka, J.; Hänggi, P. Transient anomalous diffusion in periodic systems: Ergodicity, symmetry breaking and velocity relaxation. *Sci. Rep.* **2016**, *6*, 30948. [CrossRef]
52. Kostur, M.; Machura, Ł.; Łuczka, J.; Talkner, P.; Hänggi, P. Negative conductance in driven Josephson junctions. *Acta Phys. Polon. B* **2008**, *39*, 1177.
53. Cheng, L.; Yip, N.K. The long time behavior of Brownian motion in tilted periodic potentials. *Phys. D* **2015**, *297*, 1–32. [CrossRef]

entropy

MDPI

Article

Accumulation of Particles and Formation of a Dissipative Structure in a Nonequilibrium Bath

Steven Yuvan [1,†] **and Martin Bier** [1,2,*,†]

1 Department of Physics, East Carolina University, Greenville, NC 27858, USA; yuvans16@students.ecu.edu
2 Faculty of Mechanical Engineering, Institute of Mathematics and Physics, University of Technology and Life Sciences, 85-796 Bydgoszcz, Poland
* Correspondence: bierm@ecu.edu
† These authors contributed equally to this work.

Abstract: The standard textbooks contain good explanations of how and why equilibrium thermodynamics emerges in a reservoir with particles that are subjected to Gaussian noise. However, in systems that convert or transport energy, the noise is often not Gaussian. Instead, displacements exhibit an α-stable distribution. Such noise is commonly called Lévy noise. With such noise, we see a thermodynamics that deviates from what traditional equilibrium theory stipulates. In addition, with particles that can propel themselves, so-called active particles, we find that the rules of equilibrium thermodynamics no longer apply. No general nonequilibrium thermodynamic theory is available and understanding is often ad hoc. We study a system with overdamped particles that are subjected to Lévy noise. We pick a system with a geometry that leads to concise formulae to describe the accumulation of particles in a cavity. The nonhomogeneous distribution of particles can be seen as a dissipative structure, i.e., a lower-entropy steady state that allows for throughput of energy and concurrent production of entropy. After the mechanism that maintains nonequilibrium is switched off, the relaxation back to homogeneity represents an increase in entropy and a decrease of free energy. For our setup we can analytically connect the nonequilibrium noise and active particle behavior to entropy decrease and energy buildup with simple and intuitive formulae.

Keywords: Lévy noise; nonequilibrium thermodynamics; active particles; entropy production; dissipative structures

Citation: Yuvan, S.; Bier, M. Accumulation of Particles and Formation of a Dissipative Structure in a Nonequilibrium Bath. *Entropy* **2022**, *24*, 189. https://doi.org/10.3390/e24020189

Academic Editor: Antonio M. Scarfone

Received: 17 January 2022
Accepted: 25 January 2022
Published: 27 January 2022

1. Introduction

The thermodynamics and statistical physics of particles at equilibrium is a standard part of the undergraduate curriculum. The First and Second Law of Thermodynamics are powerful concepts that lead the way to the explanation of many real-life phenomena. Further development led to notions such as the Boltzmann Distribution, the Fluctuation-Dissipation Theorem, Onsager's Reciprocal Relation, and Microscopic Reversibility [1]. Even setups that are close-to-equilibrium can often be successfully analyzed with these ideas. No general theory, however, is available for systems that are far-from-equilibrium. None of the above laws and notions apply in that case.

Imagine a liquid in which "active" particles are suspended. Such "active" particles can be bacteria that propel themselves, i.e., swim. These can also be particles that are manipulated through fields from the outside. Obviously, energy is pumped into such systems and no First Law or any of the concepts mentioned in the previous paragraph applies. Over the last two decades, setups with active particles have been the subject of much experimental and theoretical research.

There are many different ways to model the movements of active particles. One can, for instance, assume that the particle has the same speed all the time and that the change of the direction of motion follows a diffusion equation [2]. The "Run-and-Tumble" model is a more discrete version of this and it is inspired by the way that *Escherichia coli* bacteria move [3]. Here the particle or bacteria covers a finite-length straight segment at a constant speed. After coming to a stop, it lingers for a moment. It "tumbles" and then picks a new random direction for the next run. There are also different ways to let the active particle interact with the wall of the reservoir in which it swims.

In our analysis below, we focus on the 2D random walk: At every timestep, a direction is picked randomly and a displacement is drawn from a zero-centered distribution (cf. Figure 1). We let the random walks happen in a confinement. Whenever the particle hits the wall, it comes to a standstill. Subsequently, it only moves away from the wall again if a random displacement makes it move inside the circular confinement.

If displacements are drawn from a zero-average Gaussian distribution, we eventually see a homogeneous distribution of particle positions over the entire domain. However, if we instead draw distances from a so-called α-stable distribution (sometimes called a Lévy-stable distribution) [4–7], a nonhomogeneous distribution develops.

The Gaussian distribution has an exponential tail, i.e., $p_2(\xi) \propto \exp\left[-\xi^2/2\sigma^2\right]$ as $\xi \to \pm\infty$. Here σ denotes the standard deviation of the Gaussian. The rapid convergence to zero of the exponential tail means that the probability to make a big jump is very small and effectively negligible. Figure 1a shows this clearly.

For an α-stable distribution, the asymptotic behavior is described by a power law:

$$p_\alpha(\xi) \propto |\xi|^{-\alpha-1} \text{ as } \xi \to \pm\infty. \tag{1}$$

Here α is the so-called stability index for which we have $0 < \alpha < 2$. For $\alpha = 2$, the Gaussian is re-obtained. The power law converges slower than the exponential. A result of this is that outliers, i.e., large "Lévy jumps", regularly occur (see Figure 1b). Ultimately, the Lévy walk resembles a run-and-tumble walk, but, following Equation (1), the Lévy jumps have no characteristic length and the average length of a Lévy jump actually diverges.

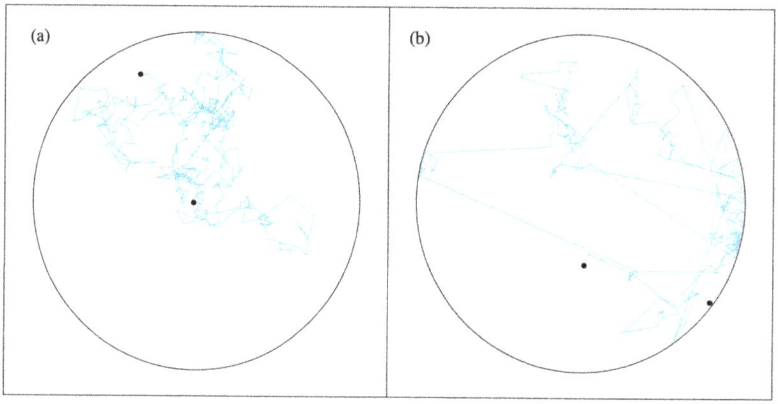

Figure 1. Random walk in a circular domain. Whenever the particle hits the wall, it comes to a standstill and later only moves again when a computed step leads to a movement inside the circle. For every step, the direction is picked randomly and the displacement is drawn from a (**a**) Gaussian distribution or from a (**b**) Lévy-stable distribution. The circle has a radius of 20. Both distributions are symmetric around zero. The Gaussian distribution has a standard deviation of $\sqrt{2}$. For the Lévy-stable distribution, we have $\alpha = 1$ and a scale factor of $\sigma = 1$.

The Central Limit Theorem [8] tells us that the Gaussian distribution is what ensues when an outcome is the result of multiple stochastic inputs. However, the theorem only applies if all of the constituent stochastic inputs have a *finite* standard deviation. For stochastic inputs with infinite standard deviations, the α-stable distribution is what results.

The α-stable distribution is a standard feature of the *Mathematica* software package and the programming for a simulation as the one leading to Figure 1b is a matter of just a few lines of code. The probability density of the α-stable distribution is given by a big and cumbersome formula [9] and we will not elaborate on it.

Alpha-stable distributions do not just provide a good model for the behavior of active particles. It turns out that power-law tails commonly occur in systems that are far-from-equilibrium with no active particles involved. Almost 60 years ago, Benoit Mandelbrot discovered that variations in the price of cotton futures follow a distribution with an $\alpha = 1.7$ power-law tail [10,11]. More power-law tails and α-stable distributions were identified in the 1990s [12–14] when desktop computers became available that could rapidly and easily perform the necessary data processing. As of yet, there is no complete and general theory to explain how and why α-stable distributions are connected to far-from-equilibrium. In this sense, the α-stable distributions are like $1/f$-noise [11,15]. The connection of far-from-equilibrium with α-stable distributions and $1/f$-noise is still for the most part, a phenomenological one.

Nevertheless, as mentioned above, nonequilibrium characteristics do emerge when, instead of Gaussian noise, Lévy noise is added to particle dynamics. Take a particle doing Brownian motion on a potential $V(x)$. Microscopic reversibility means that every trajectory $x(t)$ from an initial position x_i to a final position x_f and taking a time Δt, is traversed equally often in forward and backward direction. Microscopic reversibility is an equilibrium feature that is implied by the fact that there can be no arrow of time for a system at equilibrium, i.e., there must be time-reversal symmetry. In 1953, Onsager and Machlup gave mathematical rigor to this idea when they proved that with Gaussian noise, the most likely trajectory up a potential barrier is the reverse of the most likely trajectory down that same barrier [16–18]. It can also be rigorously proven that for Lévy noise, the most likely trajectory up a potential barrier is not the reverse of the most likely trajectory down a potential slope [19]. The presence of Lévy noise breaks the time-reversal symmetry that is implicit in equilibrium [20].

For the setup that is depicted in Figure 1b, the violation of time-reversal symmetry is in the interaction of the particle with the wall. Elastic collisions have time-reversal symmetry and had we taken the particle in Figure 1b to collide elastically with the wall, forward and backward trajectories would have been indistinguishable. Lévy jumps are rare, but because of their length, they are likely to end at the wall. Once the particle is located at the wall, the probability that the first subsequent step is already a Lévy jump away from the wall is small. Moreover, only a step that leads to a movement inside the reservoir will be processed in the simulation. Thus, the particle can "linger" near the wall after hitting it. In the end it appears as if it is easier to get to the wall than it is to get away from it, i.e., it looks as if there is reduced mobility near the wall. Figure 2 shows how this is the case on a 1D interval.

In the previous paragraph, we put the finger on something that applies generally for active particles in a confinement. They do not distribute homogeneously, but instead accumulate near a wall. It furthermore appears that the accumulation is stronger if the wall has a stronger inward curvature [21]. Active particles tend to get stuck in nooks and corners of a confinement and even more so if the nooks and corners are tighter. This is the phenomenon that we will elaborate on below.

The way Lévy particles distribute on a confined 1D segment (cf. Figure 2) can be described with a Fractional Fokker–Planck Equation [22]. The steady-state solution of that equation is available [23]. We show in Appendix A how this solution readily generalizes to higher dimensional setups. Below we examine how Lévy particles distribute over two connected reservoirs where one reservoir is a scaled down version of the other. We will see a deviation from the homogeneous distribution that is obtained when the noise is Gaussian and when equilibrium theory applies.

Suppose we have a volume V with N indistinguishable particles in it. We partition the initially empty V into two reservoirs of a volume $V/2$ each. Next the particles are inserted. Each reservoir has a probability of $1/2$ to receive each particle. Eventually, the probability for all particles to end up in one particular reservoir is 2^{-N}. The probability for an equal distribution over the two reservoirs is $\binom{N}{N/2}2^{-N}$. The binomial coefficient $\binom{N}{N/2}$ grows very rapidly with N.

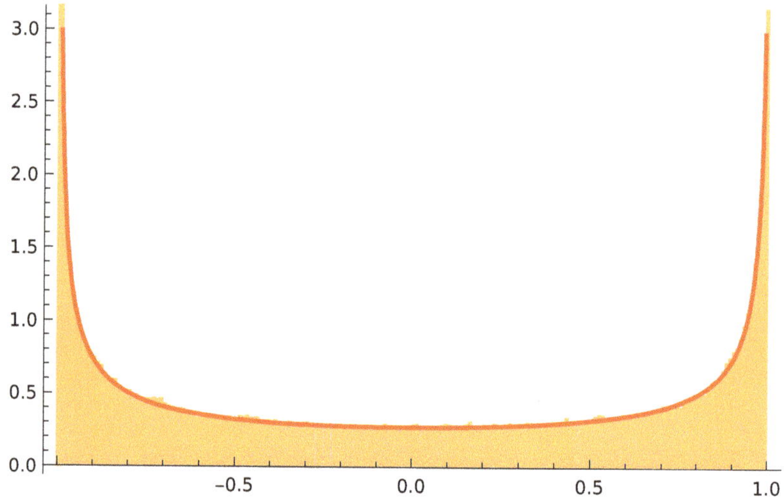

Figure 2. A Lévy walk on the interval $-1 \leq x \leq 1$ (cf. Equation (3)). The value of the stability index is $\alpha = 0.8$. Whenever the particle hits $x = \pm 1$, it stays there until an iteration occurs in the direction away from the wall. The red curve shows the analytic solution (cf. Equation (4)). The normalized histogram is the result of a numerical simulation of Equation (3); the timestep was $\Delta t = 0.001$, there were 10^7 iterations, and the scale factor of the symmetric, zero-centered Lévy distribution was taken to be one.

The reason that the air in a room never spontaneously concentrates in one half of the room is that there is just one way to put all molecules in one chosen half and $\binom{N}{N/2}$ ways to distribute them equally. In other words, the macrostate in which all air is concentrated in one particular half of the room has *one microstate* and the macrostate with a homogeneous air distribution over the entire room has $\binom{N}{N/2}$ *microstates*. The entropy of a macrostate can be defined as a scalar value that is proportional to the logarithm of the number of microstates of that macrostate [1]. In this case, it is obvious that the homogeneous distribution leads to maximal entropy.

With a partition and a pump it is, of course, possible to bring all of the air molecules to one half of the room. Such a process requires energy and with standard thermodynamics, the involved energies can be calculated. That energy-consuming, active particles can accumulate in a smaller subvolume does not violate laws of nature, and it is also possible to calculate the entropy change associated with such accumulation. We will perform such a calculation.

The ultimate goal would be a Lévy-noise-equivalent of entropy. This would be a quantity that takes its extreme value when Lévy-noise-subjected particles reach a steady state distribution. The Kullback–Leibler divergence [24] is a positive scalar value that can be thought of as a "distance" between two given distributions. The Kullback–Leibler divergence between the steady state distribution and another distribution could be a good candidate. With tools like Noether's Theorem, alternative formulations of active-particle statistical mechanics and of the Fractional Fokker–Planck Equation have been derived [25,26], with work in this direction appearing to be promising.

No general formalism is developed in this article, but we present a setup where the entropy decrease associated with the accumulation can be readily described with simple and intuitive formulae. The nonhomogeneous steady-state distributions that develop in the presence of nonequilibrium noise can be interpreted as dissipative structures [27].

The deviation from homogeneity decreases the entropy. However, active particles pump energy into the system and the dissipative structure ultimately facilitates a steady-state dissipation of energy and production of entropy.

2. The 1D and 2D Random Walk in a Confined Domain

For a particle in 1D, Brownian motion is commonly described with a Langevin equation:

$$\dot{x}(t) = \sqrt{2D}\,\xi_2(t), \quad (2)$$

where $x(t)$ is the time-dependent position of the particle, $^{\bullet} \equiv d/dt$, D is the diffusion coefficient, and $\xi_2(t)$ is a stochastic function that describes the effect of collisions with molecules in the medium. To account for the effect of such collisions, a random number $\theta_{2,i}$ is drawn at the i-th timestep. In a simulation with finite timesteps of Δt, we then take $\Delta x_i = \sqrt{2D}\,\xi_{2,i}(t_i)\,\Delta t = \sqrt{2D\Delta t}\,\theta_{2,i}$. If Δt is large enough to contain a significant number of collisions, then the aforementioned Central Limit Theorem [8] can be invoked to justify drawing the θ_i's from a zero-average Gaussian distribution. Upon taking $\langle\theta_{2,i}^2\rangle = 1$, we readily come to the traditional equation for the average squared distance that is diffused in a time interval Δt: $\langle\Delta x^2\rangle = 2D\Delta t$.

The equation $\langle\Delta x^2\rangle = 2D\Delta t$ does not contain a characteristic timescale. It is in order for the scale-free diffusion equation to ensue that we need to "adjust" the $\theta_{2,i}$'s and take $\xi_2(t_i) = \theta_{2,i}/\sqrt{\Delta t}$ in Equation (2).

In case of the 1D Lévy flight, we have for the stochastic ordinary-differential-equation and its discretized version, respectively:

$$\dot{x} = \sigma\,\xi_\alpha(t) \text{ and } \Delta x_i = \sigma\,\theta_{\alpha,i}\,\Delta t^{1/\alpha}. \quad (3)$$

Now the values for θ_α are drawn from a symmetric, zero-centered α-stable distribution with a value of one for its scale factor. The Lévy flight is still scale-free, but because $\langle\theta_{\alpha,i}^2\rangle \to \infty$ for $0 < \alpha < 2$, there is no longer a traditional diffusion equation and σ is a mere scale factor.

Figure 1 shows simulations of 2D random walks. At every timestep, a direction is chosen randomly from a flat distribution between zero and 2π. The displacement is the result of a random draw from a Gaussian distribution (Figure 1a) or from an α-stable distribution (Figure 1b). Both the Gaussian walk and Lévy walk are isotropic, i.e., taken from the center of the circle, all directions are equivalent. A generalization to more than 2 dimensions is readily formulated and simulated. The random walks then occur inside a ball with a finite radius. Whenever the domain boundary is hit, the particle comes to a standstill. For $\alpha = 2$, the random walk is symmetric under time reversal. However, as was already mentioned in the Introduction, for $0 < \alpha < 2$ the time-reversal symmetry is broken. It is not hard to understand why this is the case. When the particle is followed in forward time, we will often see a Lévy jump that makes the particle come to a standstill at the domain boundary. More rare will be a large jump from the domain boundary into the interior. When a movie of the moving particle is played backward, it will be the other way round. The forward and backward played movie are distinguishable.

Figure 2 shows the position distribution that results after a many-step 1D simulation on $-1 \leq x \leq 1$ for $\alpha = 0.8$. For $\alpha = 2$, a flat distribution results. However for $0 < \alpha < 2$, the Lévy jumps that terminate at $x = \pm 1$ and the decreased mobility there lead to an increased probability density near $x = \pm 1$. The Langevin Equation, Equation (3), can be equivalently formulated as a fractional Fokker–Planck Equation for the evolution of a probability distribution, i.e., $\partial_t p(x,t) = \sigma^\alpha \partial_x^\alpha p(x,t)$. The stationary distribution is then obtained as the solution of the ordinary differential equation that results when the left hand side is set equal to zero. The fractional derivatives are nontrivial, but in Ref. [23], a solution for the 1D case is presented:

$$p_{st}(x) = \frac{2^{1-\alpha}\Gamma(\alpha)}{\Gamma^2(\alpha/2)}\left(1 - x^2\right)^{\alpha/2-1}, \quad (4)$$

where $\Gamma(.)$ denotes the gamma function. Figure 2 shows this solution together with the results of the Langevin simulation. In Appendix A, we show with symmetry arguments that the normalized $(1-r^2)^{\alpha/2-1}$-form generalizes to the nD case, with r being the distance from the center of the ball.

It is worth noting that the U-shaped function as in Equation (4) and Figure 2 has been encountered in other systems in stochastic dynamics. For $\alpha = 1$, the Lévy stable distribution is actually the Cauchy distribution, $p_1(\xi) = (1/\pi)(\sigma/(\sigma^2 + \xi^2))$. For $\alpha = 1$ and upon taking $x = 2u - 1$, Equation (4) turns into $p(u) = (1/\pi)(u(1-u))^{-1/2}$, where $0 \leq u \leq 1$. This is called the arcsine distribution because the cumulative distribution yields an arcsine: $\int_0^u p(u')\,du' = (2/\pi)\arcsin\sqrt{u}$. In 1939, it was the same Paul Lévy who derived that the arcsine distribution emerges in the following case [28]. Let a 1D Brownian walk of duration t start at $x = 0$. Next look at the fraction of time that the Brownian particle spends on the positive semi-axis. It is found that these fractions follow an arcsine distribution. This is called the arcsine law. Recently it has been discovered that arcsine laws occur more generally [29]. Driven mesoscopic systems are obviously out-of-equilibrium, but also in such systems, an arcsine law results when one considers, for instance, fractions of time that a current stays above its average value. Arcsine laws in nonequilibrium setups is currently a much researched topic [30,31] and Equation (4) may be a manifestation of something deeper and more general.

3. Two Connected Semicircular Reservoirs

Imagine a semicircular 2D domain with radius R_1 as in Figure 3. There is a small opening with a width d that gives access to a semicircular domain with radius R_2. We have $R_2 < R_1$. Next imagine a large number of particles in this system. The particles are subjected to Lévy noise. In Appendix B, it is derived how there is a net flow into the smaller reservoir if both reservoirs have the same homogeneous particle density. Thus, when starting from thermodynamic equilibrium, a higher density develops in the smaller reservoir once Lévy noise starts being applied.

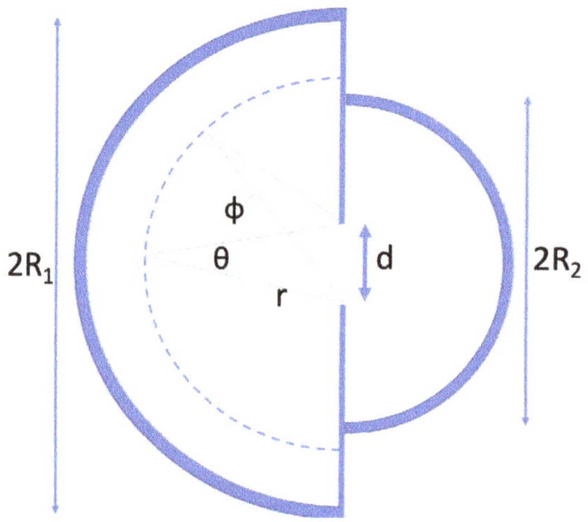

Figure 3. Two semicircular reservoirs with a small opening between them. The system contains a large number of noisy particles. At each timestep, each particle moves in an arbitrary direction with a displacement that is drawn from a Gaussian distribution or a Lévy-stable distribution as in Figure 1a,b. If a particle hits a semicircular wall, it comes to a standstill and only moves again if a computed displacement leads to motion inside the system. If a particle hits the straight vertical wall, it bounces elastically. For Gaussian noise, the system goes to an equilibrium with equal concentration on both sides of the opening. However, when the particles are subjected to Lévy noise, the steady state has an accumulation in the smaller reservoir.

3.1. Distribution over the Two Reservoirs in Case of Lévy Noise

If the noise in the setup of Figure 3 is Gaussian, then the system will relax to an equilibrium with equal concentration in the two reservoirs. Each particle then has a probability $P_1 = R_1^2/(R_1^2 + R_2^2)$ to be in the larger reservoir and a probability $P_2 = R_2^2/(R_1^2 + R_2^2)$ to be in the smaller reservoir. The probability to be in a certain reservoir is in that case, simply proportional to the volume of that reservoir. In 2D, the "volume" is the area $V_i = \pi R_i^2/2$.

Next consider Lévy particles. The distribution will now be different. As was shown in the previous section and in Appendix A, Lévy particles tend to accumulate near the walls and in the smaller "nooks and corners". With Lévy particles, the probability to be in the smaller reservoir will be larger than the reservoir's fraction of the total volume.

For a stochastic simulation, we let the semicircular walls be "sticky" again, i.e., the particle comes to a standstill upon hitting the wall and only displaces again if a subsequent computed step leads to motion inside the system. If the linear vertical wall in the middle is hit, an elastic collision occurs. Thus, that wall is "bouncy".

We will use the 2D solution for a circle, $p_{st}(r) = (\alpha/(2\pi))(1-r^2)^{\alpha/2-1}$, to come to an estimate of the steady-state distribution for the setup in Figure 3. We move to a description where $\rho_i(r_i)$, with $i = 1, 2$, denotes the normalized particle density in reservoir i at a distance r_i from the opening. With:

$$\rho_i(r_i) = \frac{\alpha}{\pi R_i^2}\left(1 - \left(\frac{r_i}{R_i}\right)^2\right)^{\alpha/2-1} \tag{5}$$

it is readily verified that:

$$\int_{r_i=0}^{R_i} \int_{\phi=-\pi/2}^{\pi/2} \rho_i(r_i) r_i \, dr_i \, d\phi = 1. \tag{6}$$

With a large number of particles in the setup, there will be a relaxation to a distribution with a fraction φ_1 in reservoir 1 and a fraction φ_2 in reservoir 2. Obviously we have $\varphi_1 + \varphi_2 = 1$. For any distribution over the two reservoirs we have:

$$\rho(r_1, r_2) = \varphi_1 \rho_1(r_1) + \varphi_2 \rho_2(r_2). \tag{7}$$

It is easy to see that $\int\int \rho(r_1, r_2) = 1$, where the integration is over the entire 2-semicircle system in the figure.

The steady state occurs if there are as many $1 \to 2$ transitions as there are $2 \to 1$ transitions. We will next derive what values of φ_1 and φ_2 lead to steady state. In the above figure, imagine a semicircular strip of width dr_i at a distance r_i from the opening. The number of particles in the strip is $\rho_i(r_i)\pi r_i dr_i$ ($i = 1, 2$). We assume that for $r > r_0$, we are in the region where the power-law-description of the tail of the Lévy distribution ($p_\alpha(r) \propto r^{-(\alpha+1)}$ as $r \to \infty$) applies. The probability that the displacement during one timestep is *larger than* r is then proportional to $r^{-\alpha}$. For small d and sufficiently large r_0, the angle θ, cf. Figure 3, will be small and we have $d = \theta r$. For a Lévy jump to lead to a particle transiting to the other reservoir, the jump must also be in the right direction. This brings in a factor $(d/r)\cos\phi$, where ϕ is the indicated angle of the position on the semicircle with the horizontal. Integrating over ϕ from $-\pi/2$ to $\pi/2$, the full direction factor is found to be $2d/r$. All in all, during one timestep we have for the number of cross-reservoir transitions from a distance between r and dr:

$$dn_i^{tr}(r_i, r_i + dr_i) \propto \varphi_i \frac{d}{r_i} r_i^{-\alpha} \rho_i(r_i) r_i \, dr_i. \tag{8}$$

Integrating from r_0 to the boundary R_i, we obtain for the number of Lévy-jump-associated transitions from reservoir i:

$$N_i^{tr} \propto \frac{\varphi_i}{R_i^2} \int_{r_i=r_0}^{R_i} r_i^{-\alpha} \left(1 - \left(\frac{r_i}{R_i}\right)^2\right)^{\alpha/2-1} dr_i. \tag{9}$$

The proportionality constant (associated with the $\propto$) and the r_0 (the radius from which the power law is taken to describe the Lévy-stable distribution) are the same for both reservoirs. At this point, it is also important to realize that for the Lévy jumps to dominate the number of $1 \rightarrow 2$ and $2 \rightarrow 1$ transitions, R_1 and R_2 must both be much larger than r_0.

Mathematica will readily give an analytical solution for the integral Equation (9). The solution involves the hypergeometric function [32]. Before working out Equation (9) in its full generality, we make a further simplification that will not affect the solution too much: As $R_i \gg r_0$ for both $i = 1$ and $i = 2$, we take $r_0 = 0$ to be the lower limit of the integral. With $0 < \alpha < 1$, the integral will not diverge with $r_i \rightarrow 0$. Next, the all-important reservoir radius R_i can be scaled out of the actual integral and incorporated in the prefactor:

$$N_i^{tr} \propto \frac{\varphi_i}{R_i^2} R_i^{-\alpha} R_i \int_{r_i=0}^{R_i} \left(\frac{r_i}{R_i}\right)^{-\alpha} \left(1 - \left(\frac{r_i}{R_i}\right)^2\right)^{\alpha/2-1} d\left(\frac{r_i}{R_i}\right). \tag{10}$$

Upon taking $u = r_i/R_i$ and $v = u^2$ (so $dv = du^2 = 2u\,du$ and thus $du = 1/(2\sqrt{v})\,dv$), further simplification is achieved:

$$N_i^{tr} \propto \varphi_i R_i^{-1-\alpha} \int_{v=0}^{1} v^{-\alpha/2-1/2}(1-v)^{\alpha/2-1}\,dv. \tag{11}$$

The integral on the right-hand side is the well-known Euler integral, which is also known as the beta function [32]. Ultimately, this integral depends only on α. It is finite for $0 < \alpha < 1$ and as it is the same for both reservoirs, we find:

$$N_i^{tr} \propto \varphi_i R_i^{-1-\alpha}. \tag{12}$$

The steady state condition is $\varphi_1 R_1^{-1-\alpha} \approx \varphi_2 R_2^{-1-\alpha}$. With $\varphi_1 + \varphi_2 = 1$ we then get:

$$\varphi_1 \approx \frac{R_1^{1+\alpha}}{R_1^{1+\alpha} + R_2^{1+\alpha}}, \quad \varphi_2 \approx \frac{R_2^{1+\alpha}}{R_1^{1+\alpha} + R_2^{1+\alpha}}, \quad \text{and} \quad \frac{\varphi_1}{\varphi_2} \approx \left(\frac{R_1}{R_2}\right)^{1+\alpha}. \tag{13}$$

The better approximation is obtained by not fully carrying through the $r_0 = 0$ simplification of the last paragraph. That the simple approximation according to Equation (13) fails for larger values of α is partly due to scaling issues. For the analytic approximation to be consistent with the numerics, we need $\Delta t^{1/\alpha}$ to be significantly smaller than r_0 (cf. Equation (3) with $\sigma = 1$). Setting $r_0 = 0$ leads to a range where this is no longer true. As α becomes larger, this range becomes larger. Keeping $r_0 > 0$ in Equation (9), we find after some algebra and use of *Mathematica* for the equivalent expression of Equation (12):

$$N_i^{tr} \propto \varphi_i \left[R_i^{-1-\alpha} - \frac{\sqrt{\pi}}{\Gamma(\frac{\alpha}{2})\Gamma(\frac{3-\alpha}{2})} \frac{r_0^{1-\alpha}}{R_i^2} \, {}_2F_1\left(\frac{1-\alpha}{2}, 1-\frac{\alpha}{2}; \frac{3-\alpha}{2}; \left(\frac{r_0}{R_i}\right)^2\right)\right], \tag{14}$$

where ${}_2F_1(a,b;c;z)$ is the aforementioned hypergeometric function. It is readily verified that the second term in the square brackets dominates for $\alpha \rightarrow 2$ and small r_0. This is due to the $r_0^{1-\alpha}$ term. The hypergeometric function is defined as a power series [32] and under the

$r_0 \ll 1$ condition we can still take $(r_0/R_i)^2 \sim 0$ and hence ${}_2F_1(.) \approx 1$. The ratio of particles in the two reservoirs is then:

$$\frac{\varphi_1}{\varphi_2} \approx \left(\frac{R_1}{R_2}\right)^{1+\alpha} \frac{\sqrt{\pi}\, R_2^{\alpha-1} - r_0^{\alpha-1}\Gamma(\frac{3-\alpha}{2})\Gamma(\frac{\alpha}{2})}{\sqrt{\pi}\, R_1^{\alpha-1} - r_0^{\alpha-1}\Gamma(\frac{3-\alpha}{2})\Gamma(\frac{\alpha}{2})}, \tag{15}$$

which reduces to $(R_1/R_2)^{1+\alpha}$ (cf. Equation (13)) if we take $\alpha < 1$ and $r_0 \to 0$ concurrently. Note, furthermore, that the equilibrium distribution, i.e., $\varphi_1/\varphi_2 = (R_1/R_2)^2$, is properly approached if we concurrently take $\alpha \to 2$ and $r_0 \to 0$. Both the approximations according to Equations (13) and (15) are depicted in Figure 4 and compared with the results of a stochastic simulation. Finally, it is worth pointing out that Equation (15) is still an approximation. The power law, Equation (1), that characterizes the Lévy-stable distribution is not valid for small values of ξ. For values of ξ near zero, the distribution is Gaussian-like and this is what is relevant for the behavior of particles close to the opening, i.e., $r \to 0$. Gaussian diffusion near the opening will lead to a continuous and differentiable steady-state concentration profile near the opening. This is also what Figure 5 shows.

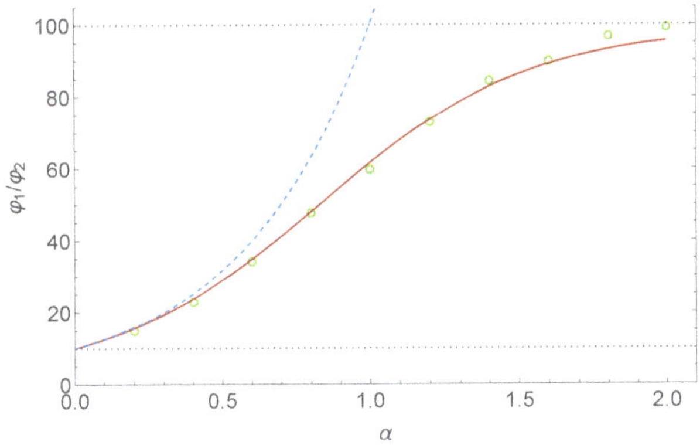

Figure 4. For the setup of Figure 3 with $R_1 = 10$ and $R_2 = 1$, we let φ_1 and φ_2 represent the fraction of particles in reservoir 1 and 2, respectively, at steady state. The curves depict the analytic approximations, Equation (13) (dashed) and Equation (15) (solid), of φ_1/φ_2. Each dot is the result of a stochastic simulation of 40,000 particles for 4×10^5 timesteps (with $\Delta t = 0.001$) following a 2×10^5 timestep relaxation period. For the approximation according to Equation (15), we let $r_0 = 0.05$ and find good agreement with the result of the stochastic simulation.

Figure 4 shows the ratio φ_1/φ_2 as a function of α and compares the result of a stochastic simulation with Equations (13) and (15). We took $R_1 = 10$ and $R_2 = 100$. For $\alpha \to 0$, the simple approximation according to Equation (13) leads to $\varphi_1/\varphi_2 = 10$. For the more sophisticated approximation according to Equation (15), the φ_1/φ_2 value at $\alpha \to 0$ can be brought arbitrarily close to 10 by taking R_1 and R_2 much larger than r_0. There is 10 times as much "sticky wall" in the large reservoir and this result tells us that for $\alpha \to 0$, all particles are concentrated at the sticky walls as would intuitively be expected.

The result that is derived in Appendix B hints at the reason that $\alpha = 1$ is "almost like" $\alpha = 2$. As we move away from the opening, the probability to hit the opening *decreases* as $r^{-\alpha}$. However, with a homogeneous distribution of particles, the number of particles at a distance between r and $r + dr$ *increases* proportional to r. For an n-dimensional setup, the increase is proportional to r^{n-1} (for $n = 2$ we have circular strips and for $n = 3$ we have spherical shells). All in all, we find that the number of "hits" from a distance r is proportional to $r^{n-\alpha-1}$. Note that for $n = 3$, the entire range of α leads to an increase of "hits" with r. We have not done any further investigation of the 3D case. We see that for $n = 2$, an increase of "hits" with r only occurs if $\alpha < 1$. For $1 < \alpha < 2$, the number of "hits" decreases with r and in that case transitions mostly happen from the region around the opening. This decrease with $1 < \alpha < 2$ also means that the particle exchange through the opening does not "feel" the different radii of the different reservoirs anymore.

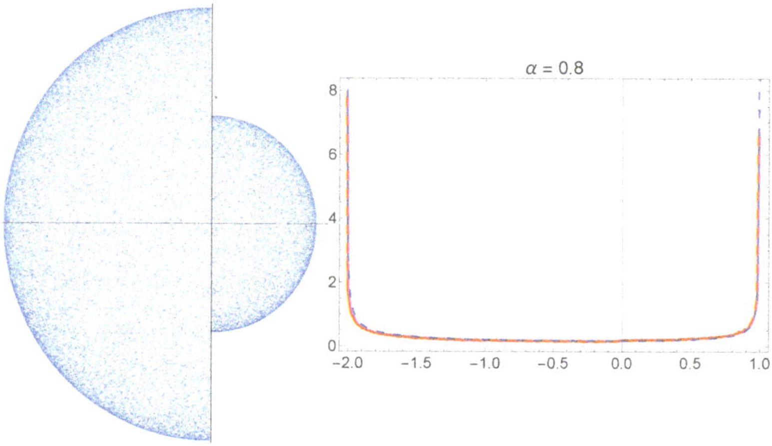

Figure 5. The figure on the left depicts a steady-state distribution for 50,000 Lévy particles in a two-reservoir confinement as depicted in Figure 3 after 10^5 timesteps. We have $R_1 = 2$, $R_2 = 1$, and the opening has a width $d = 0.1$. For the figure on the right we started with a steady-state distribution and ran the simulation for another 10^5 iterations. We took a horizontal strip through the center with a width of 0.02 and partitioned it into 300 bins. Particles in each bin were counted and the results of the subsequent 10^5 iterations were added. The solid line represents the resulting normalized 1D histogram. The dashed reference curve is the solution Equation (4). For the left reservoir, the domain was scaled to a length 2. Normalization of the combination of analytic solutions was done such that the probability to be in the left reservoir is $2/3$. It is readily verified that this leads to continuity at the location of the opening.

Equation (4) describes and Figure 5 shows a nonhomogeneous distribution: As we move away from the opening, the concentration actually increases. This should add to the exponent $n - \alpha - 1$ that we derived in a previous paragraph. Some of this effect is incorporated in the approximation that led to Equation (15). Both that approximation and the simulations show an asymptotic approach to $(R_1/R_2)^2$ as $\alpha \to 2$ and $r_0 \to 0$.

3.2. Entropies and Energies Associated with Lévy Noise

The nonhomogeneous distributions shown in Figures 2 and 5 essentially function as dissipative structures [27]. The depicted nonhomogeneous steady-state distributions represent a lower entropy than homogeneous distributions. However, these lower-entropy structures facilitate the transfer and dissipation of energy at steady state. The transferred energy comes in through the non-thermal motion of the active particles. It is next dissipated and released. Ultimately the low-entropy dissipative structures help the energy throughput and the entropy production.

As a result of the divergent standard deviation of the α-stable noise, the energy that is dissipated per unit of time is in principle infinite. The finite container size, however, truncates the Lévy jumps and make the aforementioned standard deviation of the jump sizes finite. We will not elaborate on this. What we will instead focus on in this subsection is the entropy decrease that is associated with the apparent nonhomogeneous distribution shown in Figure 5.

Imagine that the steady flow of energy that maintains the dissipative structure is suddenly halted. Such halting is straightforward if the active-particle-motion is, for instance, driven by magnetic forces or by optics. The distribution in Figure 5 will then homogenize.

Such homogenization implies an increase in entropy and a concurrent decrease in free energy. Below we will find remarkably concise analytic expressions for the entropy change.

The relaxation towards homogeneity is two-part. First there is an intra-reservoir relaxation inside each of the two reservoirs to a spatially homogeneous spread. Next there is the slower relaxation between the two reservoirs towards a ratio $\varphi_1/\varphi_2 = V_1/V_2 = R_1^2/R_2^2$.

The entropy change associated with the intra-reservoir relaxations is hard to compute for the semicircular reservoirs of Figures 3 and 5. However, for a circular reservoir as in Figure 1, it is easier and no resort to numerics is necessary. We take $p_{ini}(r) = (\alpha/2\pi)(1-r^2)^{\alpha/2-1}$ as the initial distribution and $p_{fin}(r) = 1/\pi$ as the final homogeneous distribution. It is well known that for a discrete set of probabilities, p_i, the associated entropy is given by $S = -\Sigma_i p_i \log p_i$. However, this summation cannot be straightforwardly extended to an integral for the case of a continuous probability density $p(r)$. An obvious issue is that density is not dimensionless and that a logarithm can only be taken of a dimensionless quantity. In Ref. [33], it is explained how a sensible definition is only obtained after introducing another probability density that functions as a measure. We then obtain what is known as the relative entropy or Kullback–Leibler divergence [24]:

$$D_{\mathrm{KL}}(p_{fin}||p_{ini}) = \int_{r\leq 1} p_{fin}(r)\log\left(\frac{p_{fin}(r)}{p_{ini}(r)}\right)dr. \tag{16}$$

When working out this integral, it is important to realize that the integration is from $r = 0$ to $r = 1$ over the area of a circle and that a term $2\pi r$ needs also be included. With the above expressions for $p_{ini}(r)$ and $p_{fin}(r)$, we find after some algebra that $D_{\mathrm{KL}}(p_{fin}||p_{ini}) = -1 + \alpha/2 + \log(\alpha/2)$. No such easy analytic solution ensues for more than two dimensions or even in the 1D case. The Kullback–Leibler divergence can be thought of as a kind of distance between two probability densities. However, it is generally not symmetric in the two involved distributions. In our case, we find $D_{\mathrm{KL}}(p_{ini}||p_{fin}) = -1 + 2/\alpha + \log(2/\alpha)$. Both $D_{\mathrm{KL}}(p_{fin}||p_{ini})$ and $D_{\mathrm{KL}}(p_{ini}||p_{fin})$ are remarkably simple expressions; they are continuous and concave up as α increases and reduce to zero for $\alpha = 2$.

The speed of the inter-reservoir relaxation depends on the size of the opening. For the small opening that is necessary for our approximations to be accurate, it will generally be slower than the intra-reservoir relaxation. For the inter-reservoir relaxation, the basic quantity is the probability to be in either of the two reservoirs. We go back to the basics to calculate what the entropy is for a given distribution over the two reservoirs.

In the Statistical Physics context, entropy is commonly defined as proportional to the logarithm of the number of microstates [1]. Imagine that there are N identical particles in the setup of Figures 3 and 5. Here N is taken to be very large. In case of equilibrium, the number of particles in a reservoir is proportional to the volume $V_i = \pi R_i^2/2$ of a reservoir. With $\varphi_i N$ identical particles in reservoir i, the number of microstates in each of the two reservoirs is given by:

$$\Omega_i = \frac{V_i^{\varphi_i N}}{(\varphi_i N)!}. \tag{17}$$

The numerator has the $\varphi_i N$-exponent because it is for each particle that the number of microstates is proportional to the volume. The microstate is the same, however, when two or more particles are exchanged. The denominator takes this into account and denotes the number of permutations among $\varphi_i N$ particles. With the definition $S = \log \Omega$ and using Stirling's approximation [1] ($\log N! = N \log N$, if N is very large), we derive:

$$S_i = \varphi_i N \log\left(\frac{V_i}{\varphi_i N}\right), \tag{18}$$

where "log" denotes the natural logarithm. As was mentioned before, at thermodynamic equilibrium the fraction of particles in a reservoir is proportional to the volume of that reservoir, i.e., $\varphi_i \propto V_i$. The argument of the logarithm in Equation (18) is then the same constant for both reservoirs. This leads to $S_i \propto \varphi_i$, as should be expected from an equilibrium-thermodynamics perspective.

We take for the total volume and the total entropy $V_{tot} = V_1 + V_2$ and $S_{tot} = S_1 + S_2$, respectively. It is next derived from Equation (18) that $S_{tot} = N(\varphi_1 \log(V_1/\varphi_1) + \varphi_2 \log(V_2/\varphi_2)) - N \log N$. The additive $N \log N$-term is the same for all values of α. As it is only differences in entropy that matter, we discard this term. For the entropy per particle, $s_{tot} = S_{tot}/N$, it is next found:

$$s_{tot} = \varphi_1 \log\left(\frac{V_1}{\varphi_1}\right) + \varphi_2 \log\left(\frac{V_2}{\varphi_2}\right). \tag{19}$$

Figure 6 depicts s_{tot} as a function of α following Equation (19). We took $V_{tot} = 1$ (leading to $V_1 = R_1^2/(R_1^2 + R_2^2)$ and $V_2 = R_2^2/(R_1^2 + R_2^2)$) and $R_1 = 10R_2$. For the dashed curve, Equation (13) was used to come to the values of φ_1 and φ_2. For the solid curve the improved approximation, Equation (15) was used with $r_0 = 0.05$. The curves appear almost indistinguishably close. It is important to realize that this entropy also represents free energy. The free energy release associated with the equilibration can be obtained by multiplying the entropy (cf. Equation (19)) with the temperature. Again we emphasize that Equation (19) is related to just the inter-reservoir relaxation and does not incorporate intra-reservoir relaxation.

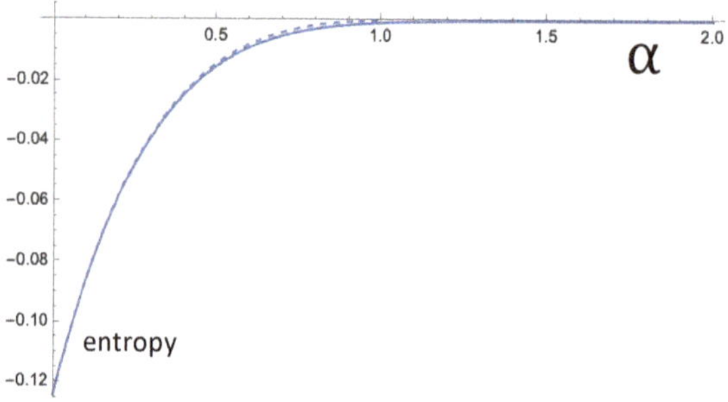

Figure 6. Given the setup of Figures 3 with $V_{tot} = 1$ and $R_1 = 10R_2$, the curves show the entropy per particle, s_{tot}, as a function of the stability parameter α of the Lévy noise. The nonequilibrium noise leads to a concentration difference between the two reservoirs. The associated entropy decrease s_{tot} is obtained by substituting into Equation (19) the approximate ratio according to Equation (13) (dashed curve) and according to Equation (15) (solid curve). For Equation (15) we took $r_0 = 0.05$, i.e., the value that led to good agreement with the stochastic stimulation (cf. Figure 4).

There is more thermodynamic way to derive the right-hand side of Equation (19) as the energy per particle that is invested in the building of the dissipative structure. With intra-reservoir equilibrium established, the chemical potential μ that is driving flux through the opening is the logarithm of the concentration ratio [1]. If we let ϕ be the fraction of the particles in the smaller reservoir, then we have $\mu(\phi) = \log\left(\frac{\phi}{V_2}\right) - \log\left(\frac{1-\phi}{V_1}\right)$. The energy that is dissipated when an infinitesimal fraction $d\phi$ follows the potential and flows through the opening is $\mu(\phi)\, d\phi$. The entire equilibration takes ϕ from φ_2 to V_2. After some algebra and setting the temperature and the Boltzmann constant all equal to unity, it is found that the resulting total-equilibration-energy integral reduces to $-s_{tot}$ (cf. Equation (19)).

Equations (13) and (19) are concise and intuitive. Equation (13) is already a fairly accurate approximation. Given the geometry of the system and the value of α, Equation (13) gives the distribution over the two reservoirs. Equation (19) tells us what entropy decrease and what free energy "investment" is associated with the concentration difference between the reservoirs that gets established due to the active particle movement. It gives a measure for how far the system is driven from equilibrium by the active particle motion.

4. Discussion

In this article we explored a significant consequence of a bath in which particles velocities are Lévy-stable distributed. With the ordinary Gaussian velocity distribution that is associated with equilibrium systems, the maximization of entropy leads to particles homogeneously distributing in the confined domain. With a Lévy-stable distribution for the velocities, larger concentrations occur near the walls and in the smaller cavities. We have analytic expressions for the distribution of Lévy particles in the circular and the spherical domain. For the two connected reservoirs as depicted in Figure 3, we have derived a good approximation for the concentration difference between the reservoirs at steady state. We have presented an accounting of the entropies, and ensuing energies, for such divergence from equilibrium.

We have interpreted the nonhomogeneous particle distribution (cf. Figure 5) as a dissipative structure, i.e., a lower-entropy arrangement of particles that facilitates a larger dissipation of energy and concurrent larger production of entropy. There is nothing about heat conduction in the equations. However, it is tempting to hypothesize that with the particles being closer to the surface area, the system would be better able to transfer heat to the environment and do so at a larger rate.

In the 1990s, experiments were performed in which DNA, RNA, and proteins were manipulated on the molecular scale. This commonly involved breaking of molecular bonds. The involved energies were significantly larger than the $k_B T$ that can be considered as the quantum of thermal energy. In such far-from-equilibrium processes, Onsager's reciprocal relations and other close-to-equilibrium concepts are no longer valid. Fortunately, at about the same time, theory was developed to handle fluctuations in far-from-equilibrium conditions. The Fluctuation Theorem [34] and the Jarzynski Equation [35,36] could very accurately account for the results of experiments in which microscopic beads were pulled by optical tweezers [37] and experiments in which RNA was forcibly unfolded [38,39]. However, it should be realized that the Fluctuation Theorem and the Jarzynski Equation apply when far-from-equilibrium events take place *in an equilibrium bath with a temperature*. In many experiments and real-life systems, the bath is the very *source of nonequilibrium*. The Fluctuation Theorem and the Jarzynski Equation are of no help in that case and new theory needs to be developed. An obstacle here is constituted by the fact that there is no equivalent of temperature for the Lévy-stable distribution of velocities that is commonly associated with the nonequilibrium bath. For a Gaussian velocity distribution, the standard deviation is proportional to the square root of the temperature. However for a Lévy-stable distribution, the standard deviation diverges and, technically, the temperature works out to be infinite. In this article we have tried to contribute to the development of description and understanding of what can happen in nonequilibrium baths.

As was explained in the Introduction, baths consisting of Lévy particles lead to similar physics as baths in which active particles are suspended. In both cases there is a continuous input of energy into the system and there is no longer a Fluctuation-Dissipation Theorem to guide the understanding and description. Swimming bacteria are a prime example of active particles. That swimming *Escherichia coli* bacteria can indeed be accumulated in cavities as has been experimentally demonstrated [40].

In a recent paper, results are presented of a numerical simulation of an active-particles-containing liquid [41]. A passive particle in this liquid was followed as a probe. This passive particle turned out to display Lévy-stable distributed displacements. What was simulated in this work was merely the Navier–Stokes equations and that passive particles exhibit these Lévy-stable distributed displacements is therefore a purely hydrodynamic effect due to active-particle-activity. That interesting and unexpected hydrodynamics can develop in liquids with immersed swimming bacteria has also been experimentally established [42].

The density profiles in Figures 2 and 5 are mindful of the coffee ring effect. When a coffee drop on a surface evaporates, the stain that is left behind is darkest towards the edge [43]. This effect is common with liquids that carry solutes. There are technological applications where it is important to control the coffee ring effect. The simple explanation for the effect goes as follows. The drop has the shape of a disk. It has a fixed radius and the height of the drop vanishes near the edge. With a uniform evaporation across the surface area of the drop, there must be a net outward radial flow to replenish lost fluid near the contact line. Solute is carried along with this flow and ultimately deposited near the contact line. Much theoretical, numerical, and experimental research has been devoted to the effect in the last quarter century (see Ref. [44] and references therein). It is common to use equilibrium concepts like Einstein's Fluctuation-Dissipation theorem when trying to account for the phenomenon. However, an evaporating drop is not in a thermodynamic equilibrium. It is certainly possible that solute particles exhibit the large jumps that are commonly encountered in nonequilibrium systems. The accumulation at the edge could then also be due to the mechanism that we discussed in this article.

In plasma physics, it is common to assume that the particles in a dense plasma follow the well-known Maxwell–Boltzmann distribution for particle speeds [1]. However, this equilibrium assumption may not always be valid, especially if the plasma is short lived and associated with an energy pulse. At Lawrence Livermore Lab, a table-top-size construction was developed to generate pulses of fast neutrons from high-energy deuterium collisions in plasma. Such collisions lead to the nuclear reaction D + D $\rightarrow^3$He + n [45]. In the experiments, it appears that the number of produced neutrons exceeds the theoretical predictions by more than an order of magnitude. The reason for this is most likely that the Maxwell–Boltzmann distribution only applies at thermodynamic equilibrium.

Plasmas in which energy is converted or transferred are of course not in a thermodynamic equilibrium. In containers with plasma, a homogeneous distribution is therefore unlikely and accumulation at the edge as described in this work is possible. This is important because it means that fusion reactions in a plasma will occur at different rates at different positions. Through feedback mechanisms, such inhomogeneities may rapidly augment and possibly develop into serious instabilities.

Engineered microswimmers is probably the field where our results could ultimately be most applicable. There are good methods and technologies for manipulating suspended micrometer size particles from the outside with acoustic, magnetic, or optic signals (see e.g., Refs. [46,47]). Today the exciting new developments are in the medical application of such microrobots. Clinical uses for imaging, sensing, targeted drug delivery, microsurgery, and artificial insemination are envisaged and researched [48]. The microswimmers and microrobots are particles that are operating in a very noisy environment. Accumulations as described and explained in this article are likely to be encountered.

Author Contributions: Conceptualization, S.Y. and M.B.; methodology, S.Y. and M.B.; software, S.Y.; formal analysis, S.Y. and M.B.; investigation, S.Y. and M.B.; writing—original draft preparation, M.B.; writing—review and editing, M.B.; supervision, M.B. All authors have read and agreed to the published version of the manuscript.

Funding: This research received no external funding.

Acknowledgments: We are grateful to Roland Winkler and Łukasz Kuśmierz for useful feedback as we were working on this article. We are also grateful to John Michael Jones and Keith Thomson for their help with the simulations that led to Figure 4.

Conflicts of Interest: The authors declare no conflict of interest.

Appendix A. Extension of the 1D Solution to *n*D

Consider the steady-state solution $p(r) = C(\alpha)\left(1 - r^2\right)^{\alpha/2-1}$, where $C(\alpha)$ is the normalization factor (cf. Equation (4)). In this Appendix we will use symmetry arguments to show that this form generalizes to higher dimensional setups.

First consider the 1D ball depicted in Figure A1a and imagine a large number of particles distributed according to Equation (4). Next take two small intervals on the right side of $r = 0$: $r_1 < r < r_1 + dr$ and $r_2 < r < r_2 + dr$, as depicted. At steady state and within any time interval Δt, there is as much flow from the r_1-interval to the r_2-interval as that there is from the r_2-interval to the r_1-interval, i.e., $J_{12} = J_{21}$. This is detailed balance [1]. Next we define a transition rate, k_{12}, that is the probability per unit of time for a particle in the r_1-interval to transit to the r_2-interval. The rate k_{21} is analogously defined. Detailed balance implies that $k_{12}p(r_1) = k_{21}p(r_2)$ and thus:

$$\frac{k_{12}}{k_{21}} = \frac{p(r_2)}{p(r_1)} = \left(\frac{1-r_2^2}{1-r_1^2}\right)^{\alpha/2-1}. \tag{A1}$$

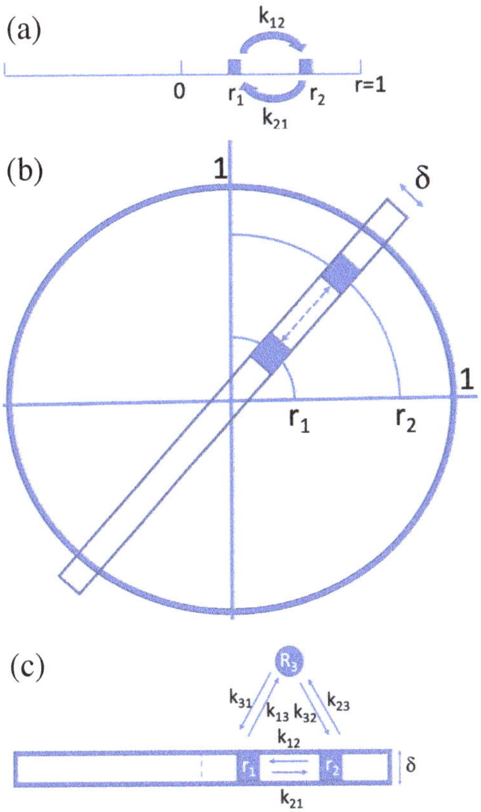

Figure A1. A Lévy walk in a confined domain. Whenever the particle hits the confinement wall, it comes to a standstill there. The 1D steady-state probability distribution (**a**) is solved in Ref. [23]: $p_{st}(r) \propto (1-r^2)^{\alpha/2-1}$. Between any two small intervals along the 1D domain, steady state implies $p(r_1)k_{12} = p(r_2)k_{21}$, where the k's denote transition rates. In 2D (**b**) there is circular symmetry. If we take any narrow bar through the origin and look exclusively at traffic inside that bar, $p_{st}(r) \propto (1-r^2)^{\alpha/2-1}$ applies again. Next, we take a state R_3 outside the bar (**c**) and include transitions between r_1 and r_2 that go via any area R_3. As the circular symmetry implies the absence of vortices, transitions k'_{12} and k'_{21} that go via R_3 must also follow $p(r_1)k'_{12} = p(r_2)k'_{21}$. From here it follows that $p_{st}(r) \propto (1-r^2)^{\alpha/2-1}$ also applies to higher dimensional setups. See the text of this Appendix for more detail.

Next consider the 2D setup depicted in Figure A1b. A bar of width δ is going through the center of the circle. We take two little areas at distances r_1 and r_2 from the center. Consider only trajectories between these two areas that stay within the bar. The traffic inside the bar should mimic the 1D system that was considered in the previous paragraph and Figure A1a. Now consider also the transitions between the two little areas that proceed through trajectories that are not restricted to the narrow bar. Without loss of generality, we take an area R_3, cf. Figure A1c, and we consider trajectories between r_1 and r_2 that pass through R_3.

It is important to realize that the circular symmetry implies that there can be no vortices within the circular domain. Flow along any simple, closed curve within the unit circle would imply that there are points with net flow in the angular direction. Thus, along the r_1, r_2, R_3-loop there must be as much clockwise flow as there is counterclockwise, i.e., $J_{cw} = J_{ccw}$. This implies $k_{12}k_{23}k_{31} = k_{13}k_{32}k_{21}$ [49] and thus:

$$\frac{k_{12}}{k_{21}} = \frac{k_{13}k_{32}}{k_{23}k_{31}}. \tag{A2}$$

The "state" R_3 can be taken to be anywhere within the circle and be of any size and shape. We can conclude that the ratio k'_{12}/k'_{21} for transitions along any path between r_1 to r_2 within the circle must be equal to the ratio k_{12}/k_{21} for transitions with trajectories inside the bar.

It follows that, for any dimensionality, the probability density at radius r must be proportional to $\left(1-r^2\right)^{\alpha/2-1}$. For a normalized probability density in n dimensions we derive:

$$p(n,r) = \frac{\Gamma\left(\frac{n+\alpha}{2}\right)}{\pi^{n/2}\Gamma\left(\frac{\alpha}{2}\right)}\left(1-r^2\right)^{\alpha/2-1}. \tag{A3}$$

For $n = 2$ the prefactor reduces to a simple $\alpha/(2\pi)$.

Appendix B. Traffic between the Larger and the Smaller Semicircular Reservoir

Consider the setup in Figures 3 and 5 and let a homogeneous particle density ρ be same in both reservoirs. In this Appendix we show that, if the particles are subject to Lévy noise, accumulation in the smaller reservoir develops. Imagine a semicircular strip of width dr at a distance r from the opening. There are $\rho\pi r dr$ particles in this strip. We let the power-law approximation, cf. Equation (1), be valid for $r > r_0$. Here r_0 is much larger than the width of the opening d and much smaller than the radii R_1 and R_2. The angle θ is small and with θ expressed in radians we have $d = \theta r$. For a particle in the semicircular strip at distance $r > r_0$, there is a probability that a Lévy jump will bring it to the other reservoir. To achieve such transition, the jump needs to be larger than r. For such a jump, the probability is proportional to $r^{-\alpha}$. In order to go through the opening, the jump must also be in the right direction. This leads to a factor $(d/r)\cos\phi$ (cf. Figure 3). After the integration from $\phi = -\pi/2$ to $\phi = \pi/2$, we derive a "direction factor" of $2d/r$, i.e., $\propto 1/r$. Putting together all of the effects specified in this paragraph, we have the following formula for the number of transitions during a small timestep from a distance between r and $r + dr$:

$$dn^{tr}(r, r+dr) \propto \frac{1}{r} r^{-\alpha} r\, dr = r^{-\alpha}\, dr. \tag{A4}$$

Next we integrate from r_0 to the boundary R_i ($i = 1, 2$) and find for the number of Lévy-jump-associated transitions from reservoir i:

$$N_i^{tr} \propto \int_{r_0}^{R_i} r^{-\alpha}\, dr \propto \mathrm{sgn}(1-\alpha)\left(R_i^{1-\alpha} - r_0^{1-\alpha}\right). \tag{A5}$$

Care must be taken in case of $\alpha = 1$. In that case $N_i^{tr,\alpha=1} \propto \log R_i - \log r_0$. We thus conclude that for $0 < \alpha \leq 1$, the number N_i^{tr} diverges as R_i is taken to infinity. For $1 < \alpha < 2$, a constant value for N_i^{tr} ensues if $R_i \to \infty$.

The proportionality constant (associated with the $\propto$) and r_0 (the radius from which the power law is taken to describe the Lévy distribution) are the same for both reservoirs.

Thus, if both reservoirs in Figures 3 and 5 have the same uniform ρ, then we find for the net number of particles $\Delta N^{tr} = N_1^{tr} - N_2^{tr}$ that transits from the larger to the smaller reservoir:

$$\Delta N^{tr} \propto \mathrm{sgn}(1-\alpha)\left(R_1^{1-\alpha} - R_2^{1-\alpha}\right). \tag{A6}$$

If $0 < \alpha \leq 1$ and if values for R_1 and R_2 are large, then there is accumulation in the smaller reservoir.

For $1 < \alpha < 2$, there will again be accumulation in the smaller reservoir, but the effect becomes smaller as R_2 and R_1 grow and will become negligible as $R_{1,2} \to \infty$. Effectively, the geometry of the reservoirs is irrelevant for large R_1 and R_2. In that case it is particles near the opening that dominate the traffic through the opening.

In the main text, the above derivation is carried out for the case of a density $\rho_i(r_i)$ $(i = 1, 2)$ that depends on the distance r_i from the opening.

Finally, it is worth pointing out that the above derivation readily generalizes to higher dimensional reservoirs. In the 3D case, we face hemispheres. The number of particles in a hemispheric shell is $\rho 2\pi r^2 dr$. For the nD case, the shell contains a number of particles that is proportional to $r^{n-1}dr$. We thus have for dn_{nD}^{tr}:

$$dn_{nD}^{tr}(r, r+dr) \propto \frac{1}{r} r^{-\alpha} r^{n-1}\, dr = r^{n-2-\alpha}\, dr. \tag{A7}$$

This leads to:

$$N_{i,nD}^{tr} \propto \int_{r_0}^{R_i} r^{n-2-\alpha}\, dr \propto \left(R_i^{n-1-\alpha} - r_0^{n-1-\alpha}\right). \tag{A8}$$

and

$$\Delta N_{nD}^{tr} \propto \left(R_1^{n-1-\alpha} - R_2^{n-1-\alpha}\right). \tag{A9}$$

This is an interesting result. For 3 and more dimensions, we do not need to discriminate between different ranges of α. Lévy noise with any α $(0 < \alpha < 2)$ will in that case lead to a significantly higher density in the smaller reservoir and the effect will be stronger for higher values of $R_{1,2}$.

References

1. Moore, W.J. *Physical Chemistry*; Longman: London, UK, 1972.
2. Das, S.; Gompper, G.; Winkler, R.G. Confined active Brownian particles: Theoretical description of propulsion-induced accumulation. *New J. Phys.* **2018**, *20*, 015001. [CrossRef]
3. Purcell, E.M. Life at low Reynolds number. *Am. J. Phys.* **1977**, *45*, 3–11. [CrossRef]
4. Klafter, J.; Shlesinger, M.F.; Zumofen, G. Beyond Brownian motion. *Phys. Today* **1996**, *49*, 33–39. [CrossRef]
5. Lévy, P. *Calcul des Probabilités*; Gauthier-Vollars: Paris, France, 1925.
6. Zolotarev, V.M. *One-Dimensional Stable Distributions*; American Mathematical Society: Providence, RI, USA, 1986.
7. Samorodnitsky, G.; Taqqu, M.S. *Stable Non-Gaussian Random Processes*; Chapman & Hall: London, UK, 1994.
8. Van Kampen, N.G. *Stochastic Processes in Physics and Chemistry*; Elsevier: Amsterdam, The Netherlands, 1992.
9. Górska, K.; Penson, K.A. Lévy stable two-sided distributions: Exact and explicit densities for asymmetric case. *Phys. Rev. E* **2011**, *83*, 061125. [CrossRef]
10. Mandelbrot, B. The variation of certain speculative prices. *J. Bus.* **1963**, *36*, 394–419. [CrossRef]
11. Mandelbrot, B. *The Fractal Geometry of Nature*; W.H. Freeman: San Francisco, CA, USA, 1983.
12. Peng, C.-K.; Mietus, J.; Hausdorff, J.M.; Havlin, S.; Stanley, H.E.; Goldberger, A.L. Long-range anticorrelations and non-Gaussian behavior of the heartbeat. *Phys. Rev. Lett.* **1993**, *70*, 1343–1346. [CrossRef] [PubMed]
13. Penna, T.J.P.; de Oliveira, P.M.C.; Sartorelli, J.C.; Gonçalves, W.M.; Pinto, R.D. Long-range anticorrelations and non-Gaussian behavior of a leaky faucet. *Phys. Rev. E* **1995**, *52*, R2168–R2171. [CrossRef]
14. Boffetta, G.; Carbone, V.; Giuliani, P.; Veltri, P.; Vulpiani, A. Power laws in solar flares: Self-organized criticality or turbulence? *Phys. Rev. Lett.* **1999**, *83*, 4662–4665. [CrossRef]
15. Milotti, E. 1/f noise: A pedagogical review. *arXiv* **2002**, arXiv:physics/0204033.
16. Onsager, L.; Machlup, S. Fluctuations and irreversible processes. *Phys. Rev.* **1953**, *91*, 1505–1512. [CrossRef]
17. Bier, M.; Astumian, R.D. What is adiabaticity?—Suggestions from a fluctuating linear potential. *Phys. Lett. A* **1998**, *247*, 385–390. [CrossRef]

18. Bier, M.; Derényi, I.; Kostur, M.; Astumian, R.D. Intrawell relaxation of overdamped Brownian particles. *Phys. Rev. E* **1999**, *59*, 6422–6432. [CrossRef] [PubMed]
19. Kuśmierz, Ł.; Chechkin, A.V.; Gudowska-Nowak, E.; Bier, M. Breaking microscopic reversibility with Lévy flights. *Eur. Lett.* **2016**, *114*, 60009. [CrossRef]
20. Yuvan, S.; Bier, M. The breaking of time-reversal symmetry for a particle in a parabolic potential that is subjected to Lévy noise—Theory and an application to solar flare data. *Phys. Rev. E* **2021**, *104*, 014119. [CrossRef]
21. Fily, Y.; Baskaran, A.; Hagan, M.F. Dynamics of self-propelled particles under strong confinement. *Soft Matter* **2014**, *10*, 5609–5617. [CrossRef]
22. Jespersen, S.; Metzler, R.; Fogedby, H.C. Lévy flights in external force fields: Langevin and fractional Fokker-Planck equations and their solutions. *Phys. Rev. E* **1999**, *59*, 2736–2745. [CrossRef]
23. Denisov, S.I.; Horsthemke, W.; Hänggi, P. Steady-state Lévy flights in a confined domain. *Phys. Rev. E* **2008**, *77*, 061112. [CrossRef]
24. Kullback, S.; Leibler, R.A. On information and sufficiency. *Ann. Math. Statist.* **1951**, *22*, 79–86. [CrossRef]
25. Kadkhoda, N.; Lashkarian, E.; Inc, M.; Akinlar, M.A.; Chu, Y.-M. New exact solutions and conservation laws to the fractional-order Fokker–Planck equations. *Symmetry* **2020**, *12*, 1282. [CrossRef]
26. Hermann, S.; Schmidt, M. Noether's theorem in statistical mechanics. *Commun. Phys.* **2021**, *4*, 176. [CrossRef]
27. Kondepudi, D.; Prigogine, I. *Modern Thermodynamics: From Heat Engines to Dissipative Structures*; Wiley & Sons: New York, NY, USA, 1998.
28. Lévy, P. Sur certains processus stochastiques homogènes. *Compos. Math.* **1939**, *7*, 283–339.
29. Barato, A.C.; Roldán, E.; Martínez, I.A.; Pigolotti, S. Arcsine laws in stochastic thermodynamics. *Phys. Rev. Lett.* **2018**, *121*, 090601. [CrossRef] [PubMed]
30. Wang, W.; Höll, M.; Barkai, E. Large deviations of the ballistic Lévy walk model. *Phys. Rev. E* **2020**, *102*, 052115. [CrossRef] [PubMed]
31. Spiechowicz, J.; Łuczka, J. Arcsine law and multistable Brownian dynamics in a tilted periodic potential. *Phys. Rev. E* **2021**, *104*, 024132. [CrossRef]
32. Abramowitz, M.; Stegun, I.A. *Handbook of Mathematical Functions: With Formulas, Graphs, and Mathematical Tables*; Dover Publications: New York, NY, USA, 1965.
33. Jaynes, E.T. Information theory and statistical mechanics. In *Statistical Physics—1962 Brandeis Summer Institute Lectures in Theoretical Physics*; Ford, K., Ed.; Benjamin: New York, NY, USA, 1963; Volume 3, pp. 181–218.
34. Evans, D.J.; Searles, D.J. Equilibrium microstates which generate second law violating steady states. *Phys. Rev. E* **1994**, *50*, 1645–1648. [CrossRef]
35. Jarzynski, C. Nonequilibrium equality for free energy differences. *Phys. Rev. Lett.* **1997**, *78*, 2690–2693. [CrossRef]
36. Jarzynski, C. Equilibrium free-energy differences from nonequilibrium measurements: A master-equation approach. *Phys. Rev. E* **1997**, *56*, 5018–5035. [CrossRef]
37. Wang, G.M.; Sevick, E.M.; Mittag, E.; Searles, D.J.; Evans, D.J. Experimental demonstration of violations of the second law of thermodynamics for small systems and short time scales. *Phys. Rev. Lett.* **2002**, *89*, 050601. [CrossRef]
38. Liphardt, J.; Dumont, S.; Smith, S.B.; Tinoco, I., Jr.; Bustamante, C. Equilibrium information from nonequilibrium measurements in an experimental test of Jarzynski's equality. *Science* **2002**, *296*, 1832–1835. [CrossRef]
39. Collin, D.; Ritort, F.; Jarzynski, C.; Smith, S.B.; Tinoco, I.; Bustamante, C. Verification of the Crooks fluctuation theorem and recovery of RNA folding free energies. *Nature* **2005**, *437*, 231–234. [CrossRef]
40. Galajda, P.; Keymer, J.; Chaikin, P.; Austin, R. A wall of funnels concentrates swimming bacteria. *J. Bacteriol.* **2007**, *189*, 8704–8707. [CrossRef] [PubMed]
41. Kanazawa, K.; Sano, T.G.; Cairoli, A.; Baule, A. Loopy Lévy flights enhance tracer diffusion in active suspensions. *Nature* **2020**, *579*, 364–367. [CrossRef] [PubMed]
42. Sipos, O.; Nagy, K.; Di Leonardo, R.; Galajda, P. Hydrodynamic trapping of swimming bacteria by convex walls. *Phys. Rev. Lett.* **2015**, *114*, 258104. [CrossRef] [PubMed]
43. Deegan, R.D.; Bakajin, O.; Dupont, T.F.; Huber, G.; Nagel, S.R.; Witten, T.A. Capillary flow as the cause of ring stains from dried liquid drops. *Nature* **1997**, *389*, 827–829. [CrossRef]
44. Moore, M.R.; Vella, D.; Oliver, J.M. The nascent coffee ring: How solute diffusion counters advection. *J. Fluid Mech.* **2021**, *920*, A54. [CrossRef]
45. Zweiback, J.; Cowan, T.E.; Smith, R.A.; Hartley, J.H.; Howell, R.; Steinke, C.A.; Hays, G.; Wharton, K.B.; Crane, J.K.; Ditmire, T. Characterization of fusion burn time in exploding deuterium cluster plasmas. *Phys. Rev. Lett.* **2000**, *85*, 3640–3643. [CrossRef]
46. Jiang, H.-R.; Yoshinaga, N.; Sano, M. Active motion of a Janus particle by self-thermophoresis in a defocused laser beam. *Phys. Rev. Lett.* **2010**, *105*, 268302. [CrossRef]
47. Peng, X.; Chen, Z.; Kollipara, P.S.; Liu, Y.; Fang, J.; Lin, L.; Zheng, Y. Opto-thermoelectric microswimmers. *Light. Sci. Appl.* **2020**, *9*, 141. [CrossRef]
48. Bunea, A.-I.; Taboryski, R. Recent advances in microswimmers for biomedical applications. *Micromachines* **2020**, *11*, 1048. [CrossRef]
49. Hill, T.L. *Free Energy Transduction and Biochemical Cycle Kinetics*; Dover Publications: New York, NY, USA, 2004.

entropy

MDPI

Communication

Spherulites: How Do They Emerge at an Onset of Nonequilibrium Kinetic-Thermodynamic and Structural Singularity Addressing Conditions?

Jacek Siódmiak * and Adam Gadomski

Institute of Mathematics and Physics (Group of Modeling of Physicochemical Processes), Faculty of Chemical Technology and Engineering, Bydgoszcz University of Science and Technology, 85-796 Bydgoszcz, Poland; agad@pbs.edu.pl
* Correspondence: siedem@pbs.edu.pl; Tel.: +48-523-408-697

Abstract: This communication addresses the question of the far-from-equilibrium growth of spherulites with different growing modes. The growth occurs in defects containing and condensed matter addressing environments of (bio)polymeric and biominerals involving outcomes. It turns out that it is possible to anticipate that, according to our considerations, there is a chance of spherulites' emergence prior to a pure diffusion-controlled (poly)crystal growth. Specifically, we have shown that the emergence factors of the two different evolution types of spherulitic growth modes, namely, diffusion-controlled growth and mass convection-controlled growth, appear. As named by us, the unimodal crystalline Mullins–Sekerka type mode of growth, characteristic of local curvatures' presence, seems to be more entropy-productive in its emerging (structural) nature than the so-named bimodal or Goldenfeld type mode of growth. In the latter, the local curvatures do not play any crucial roles. In turn, a liaison of amorphous and crystalline phases makes the system far better compromised to the thermodynamic-kinetic conditions it actually, and concurrently, follows. The dimensionless character of the modeling suggests that the system does not directly depend upon experimental details, manifesting somehow its quasi-universal, i.e., scaling addressing character.

Keywords: spherulites; (poly)crystal formation; complex growing phenomenon; soft condensed matter; nonequilibrium thermodynamics; physical kinetics; entropy production

Citation: Siódmiak, J.; Gadomski, A. Spherulites: How Do They Emerge at an Onset of Nonequilibrium Kinetic-Thermodynamic and Structural Singularity Addressing Conditions? *Entropy* **2022**, *24*, 663. https://doi.org/10.3390/e24050663

Academic Editor: Antonio M. Scarfone

Received: 12 April 2022
Accepted: 6 May 2022
Published: 9 May 2022

1. Introduction

Defects in condensed matter phase are ubiquitous in their appearances and types. As for the point defects, one can mention atomic vacancies and interstitial positions for the abundant atoms covering the corresponding material's locations.

As for extended defects, in turn, it is appropriate to list grain boundaries, two-dimensional defects, very characteristic of polycrystals, whether composed of metallic or of (bio)polymeric material. Stepping up one dimension higher, one can address (screw) dislocations and disclinations, to mention but two [1].

Spherulites are complex defects emerging in condensed phase. They bear something from the notion of extended defects, but they mean something distinctly more than this. They are typically recognized as an imperfect crystalline phase comprised from radially distributed polycrystal's needles, crossed at certain non-crystallographic angles with each other, but intermingled with a not-yet-crystallized amorphous phase in between, cf. Figure 1. The most known technique to visualize the spherulites as defects is polarized light microscopy, which yields the famous Maltese cross. Especially, the amorphous, non-radial phase of the object is named a "band defect" the expression of which depends on peculiarities of the system studied [2]. It seems as if it was a process of competition between crystal and amorphous emerging phases that tend to evolve in a diffusional manner, $r(t) \sim t^{1/2}$ (r—arbitrary spherulite's radius in diffusion-controlled mode; t—time), and then, they change their growing mode toward its mass-convection-like (long times) counterpart, $r(t) \sim t$. This implies that the evolution goes with a constant speed [3–5] and proceeds with certain massive steps in absorbing the amorphous matter of the system. It looks as if it was that the emerging (poly)crystalline phase imposes a confinement on the diffusion space, yielding an accelerated absorption of the diffusing material onto the crystal phase, with a remnant non-absorbed amorphous phase still available for diffusion to occur. The method of revealing this phenomenon turns out to be the DSC (Differential Scanning Calorimetry), and the underlying process is coalescence of the spherulitic material, also resulting in a structural impingement of the spherulites [6].

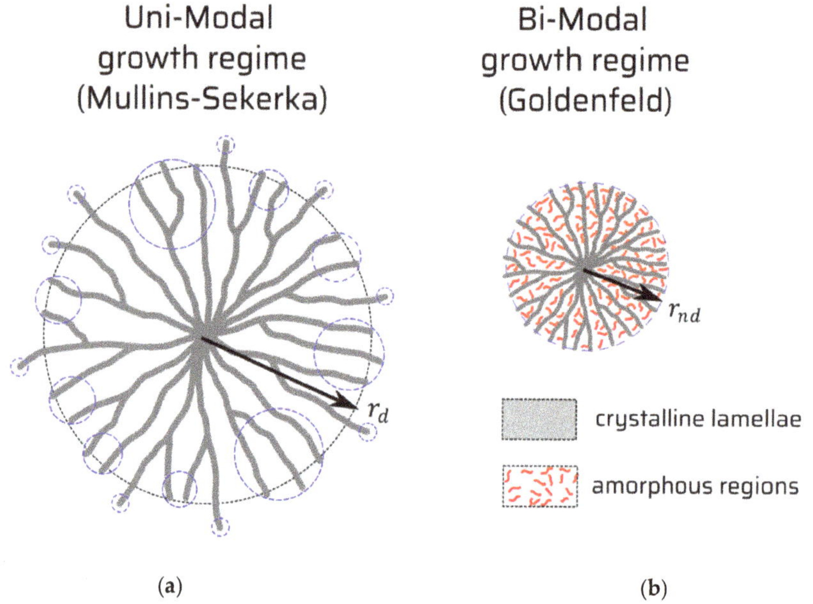

Figure 1. Mullins–Sekerka (M–S) type and Goldenfeld (G) type: (**a**) M–S, local curvatures are indicated by blue dashed circles; (**b**) G, mean curvature is signified by blue dashed circle.

An intriguing question that appears sounds: Why the spherulitic evolution changes its mode from the diffusion-like to that mass-convection-like? Amongst many answers to this question, there is at least a pronounced streamline of arguments proclaiming that the growing system of interest is evolving in nonequilibrium thermodynamic boundary conditions [3,7]. Being motivated by the aforepresented and not-answered-in-full experimental observations, in what follows, we are attempting to provide a simple theoretical rationalization that it is convincingly seen in terms of our type of modeling.

To achieve our goal, we shall employ in a natural way a spherical approximation to a conserved-mass deposition, mimicking the spherulitic growth. It seems to be really natural here, because the complex defects called spherulites assume ultimately a sphere-like form. The spherical approximation to the mass conservation law has been presented elsewhere [8–10]. However, the essential novelty applied to it is going to rely on a special boundary condition of nonequilibrium character [11]. Another relevant precondition applied to the spherulite's modeling, especially important for (bio)macromolecular realizations, is that the mass-convective instead of purely diffusive mass transportation conditions are employed decisively to create the spherulite's evolution. However, the diffusion limit is not completely ruled out, but it is also discussed when considering the onset of spherulitic formations [8–11].

In the subsequent sections, we shall present the model of spherulitic formation, capturing both diffusional and non-diffusional/mass-convection-like competitive modes, and bearing a signature of nonlinear ordinary differential equation (ode), solvable when resorting to its numerical solutions [12]. The presentation of the spherulite-formation model in terms of nonlinear ode with the corresponding initial conditions gives also a chance of its qualitative analysis. It leads to determining spherulite's characteristic linear dimension valid for both regimes recalled, albeit the radius' value for the immature spherulite (rather, its prerequisite) at the onset of the spherulitic growth is half the size as it would be for its purely diffusive counterpart. The proposed semi-quantitative model looks fairly manageable to solicit firm-basis addressing conclusions toward spherulites' formation. It is because this is to a major extent presented in terms of rescaled variables [12], both independent and dependent (the key parameters as well). Thus, all quantities of interest are nondimensional, and the number of the key governing parameters is reduced from five to two, see Section 2. It univocally allows to conclude on the principal features of the proposed modeling, showing up basic signatures of the spherulitic growth [3,4,7], and its, most importantly, inherent passage from diffusional to mass-convection-like limit, a type of structural, nonequilibrium phase transition [6].

The article is organized as follows. In Section 2, a kinetic-thermodynamic nonequilibrium model of spherulitic growth is presented, whereas in Section 3 its main results toward the onset of the spherulitic growth are disclosed, and its properties, also certain proposals for legitimate and/or useful extensions, are discussed. Section 4 serves for the main conclusions.

2. Spherulites' Formation in Terms of a Kinetic-Thermodynamic Model

Herein below, let us consider in brief a model of the spherulitic growth that is based on a mass-convection conserved field instead of a diffusion field. We would like to state clearly that, considering the growth of spherulites, here from solution of a certain concentration, we propose our simple approach in which the spherulites are represent by spherical objects. Note that a spherulite is a 3D-system (there also exist some 2D objects commonly known as cylindrulites). The non-equilibrium character of the process can quantitatively be manifested by at least: (i) external concentration field feeding the growing object, (ii) internal boundary condition prescribed at the interface: spherulite-surroundings.

It is worth recalling the following experimental observation: the growth rate $v \equiv dR/dt$ (R is the spherulite's radius; $R \equiv R(t)$, where t is time) is mainly a parametric function of temperature T (the process under study is isothermal) and slightly depends on a particular system of interest. Thus, we may solely expect that asymptotically $R \propto t$. Note that, especially in the long times' domain, it substantially differs from the well-known relationship $R \propto t^{1/2}$ characteristic for purely diffusion-controlled crystal growth processes, as first uncovered by the perennially alive Mullins–Sekerka approach [13]. This approach assumes that the growth rate depends on local curvatures of the interface, the growing object vs. surroundings. From this, the square root radius vs. time relationship emerges.

An evolution equation for spherulites can be formulated as follows:

(i) the mass conservation law as a fundamental evolution equation for growing spherulite, with an initial condition which is an initial shape (a surface) of the growing spherulite;
(ii) specification of the concentration field of the mass-feeding medium and of the spherulite at the interface; it, in general, should be associated with non-equilibrium boundary conditions;
(iii) specification of fluxes through the interface and its connection with the concentration field of the corresponding surroundings; in general, it allows to introduce not exclusively diffusive but also others, such as mass-convection fluxes of atoms, molecules, oligomers and aggregates, etc.

It has been shown that an evolution equation for growing objects (like polycrystals) with an ideal or perturbed spherical symmetry has the form:

$$[C - c(R)]\dot{R} = -\vec{j}\,[c(R)] \circ \vec{n}_0, \tag{1}$$

where $\dot{R} = dR/dt$, C is the object's density, which may depend on space variables and can generally be of stochastic nature, $c(R)$ stands for concentration of external particles at the surface, $\vec{j}\,[c(R)]$ is the flux of particles outside the object which depends functionally on concentration and $\vec{n}_0$ is the outer normal to the surface of the object. Both sides of Equation (1) are given in SI units of kg/m^2s.

As regarding point (ii), the concentration of the particles at the surface of the growing object is determined by thermodynamic conditions and geometry of the surface. Under assumption of local thermodynamical equilibrium near the interface, it has the form of the Gibbs–Thompson relation. In a more realistic model, the surface is far from equilibrium and its deviation from equilibrium is proportional to the growth velocity of the interface:

$$c(R) = c_0\left(1 + \frac{2\Gamma}{R} - \beta\dot{R}\right), \tag{2}$$

where c_0 is the concentration field at a flat interface, Γ is the capillary constant which aims at smoothening out the surface of the growing object and is proportional to the surface tension [14], β is a positive kinetic coefficient, $2/R$ is twice the mean curvature of the spherical object, and the last term describes a deviation from the thermodynamic equilibrium. When $\beta = 0$, one gets the well known Gibbs–Thomson condition.

Let us consider point (iii). If the feed of the growing object is purely mass-convective, then:

$$\vec{j}\,[c(R)] = c(R)\vec{v}\,(R), \tag{3}$$

where $\vec{v}\,(R)$ is a mass-convection velocity. (For the diffusion-limited, Mullins–Sekerka type growth, r.h.s. of Equation (3) assumes a concentration-gradient form [13]).

From the above, a basic evolution equation can be derived:

$$\left[C - c_0\left(1 + \frac{2\Gamma}{R} - \beta\dot{R}\right)\right]\dot{R} = c_0 v_0\left[1 + \frac{2\Gamma}{R} - \beta\dot{R}\right], \tag{4}$$

where $R(t = t_0) > 0$.

The growth process described by Equation (4) is influenced by five parameters: C, c_0, Γ, β and v_0. However, in fact, only two parameters are physically meaningful: (i) the quantity Δ which is a measure of the saturation in the system:

$$\Delta = \frac{C - c_0}{c_0} = \frac{C}{c_0} - 1, \tag{5}$$

and (ii) rescaled kinetic coefficient β_0:

$$\beta_0 = \beta v_0. \tag{6}$$

Indeed, rescaling the bare variables R and t to dimensionless quantities $r = r(\tau)$ and τ via the relations:

$$r = \frac{R}{2\Gamma}, \tag{7}$$

and

$$\tau = \frac{v_0}{2\Gamma}t \tag{8}$$

is useful for carrying out a solid semi-quantitative description of the spherulite's formation equation.

From the system (4)–(8), one derives the following nonlinear differential equation

$$\beta_0\left(\frac{dr}{d\tau}\right)^2+\left(\Delta+\beta_0-\frac{1}{r}\right)\frac{dr}{d\tau}-\frac{1}{r}-1=0. \tag{9}$$

Equation (9) is an algebraic quadratic equation with respect to $dr/d\tau$, where $x = dr/d\tau$, which can be rewritten as:

$$\beta_0 x^2+\frac{(\Delta+\beta_0)r-1}{r}x-\frac{r+1}{r}=0. \tag{10}$$

The real valued roots of this equation can be determined by using the conventional method of solving quadratic equations, namely, with specifying the characteristic Δ_x:

$$\Delta_x=\frac{[(\Delta+\beta_0)r-1]^2+4\beta_0 r(r+1)}{r^2}>0, \tag{11}$$

and its necessary square root:

$$\sqrt{\Delta_x}=\frac{\sqrt{[(\Delta+\beta_0)r-1]^2+4\beta_0 r(r+1)}}{r}:=d(r)/r, \tag{12}$$

thus, if the numerator of the fraction in Equation (12) is denoted for convenience by $d(r)$. The roots of the quadratic equation are explicitly given by:

$$x_{1|2}=\frac{-[(\Delta+\beta_0)r-1]\pm\sqrt{[(\Delta+\beta_0)r-1]^2+4\beta_0 r(r+1)}}{2\beta_0 r}. \tag{13}$$

One of its roots has to be ruled out. It is determined by the limiting case $\beta_0 \to 0$. (This, contrary to the vanishing kinetic limit of the phenomenon, can rather be ascribed to its thermodynamic counterpart.) Finally, one gets:

$$\frac{dr}{d\tau}=x_2=\frac{\sqrt{[(\Delta+\beta_0)r-1]^2+4\beta_0 r(r+1)}}{2\beta_0 r}+\frac{1}{2\beta_0 r}-\frac{\Delta+\beta_0}{2\beta_0}, \tag{14}$$

or by employing the shorter notation with $d(r)$

$$\frac{dr}{d\tau}=\frac{d(r)+1}{2\beta_0 r}-\frac{\Delta+\beta_0}{2\beta_0}>0, \tag{15}$$

where explicitly

$$d(r)=\sqrt{[(\Delta+\beta_0)r-1]^2+4\beta_0 r(r+1)}. \tag{16}$$

Notice that, by presenting above the rescaled spherulite's evolution equation, that means Equation (15) with its supporting Equation (16), we have arrived at the equation on which we wish to perform a simplified structural stability analysis in order to reveal the onset of the spherulitic growth.

Numerical solutions of Equation (15) by using Euler's discretization method are presented in Figure 2. The dependence of $r(\tau)$ on the rescaled kinetic (β_0) and thermodynamic (Δ) dimensionless parameters indicate the edge of diffusion- and mass-convection-driven growth at the onset of spherulitic formation around $\beta_0 \approx \Delta \approx 0.2$. A square root tendency manifests earlier for the set of upper curves, thus for smaller values of β_0.

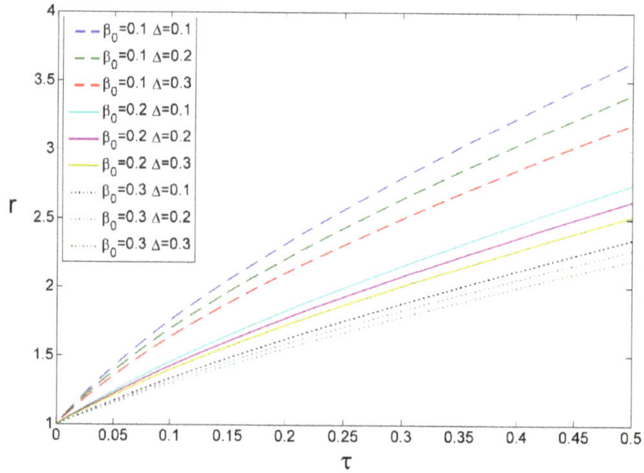

Figure 2. The dependence of $r(\tau)$ on the rescaled kinetic β_0 and thermodynamic Δ dimensionless parameters. In the chosen time interval, the curves reflect a visible tendency to pass from the diffusion-like ($\beta_0 \to 0$) to mass-convective-type mode ($\beta_0 \nrightarrow 0$). Other realizations of $r(\tau)$ are presented in [12].

3. Results and Discussion

First of all, let us note that, at the onset of the spherulitic growth at which the overall system is settled in front of a decision whether it will evolve either in diffusional or in non-diffusional (mass-convection-like) limit, one can derive the double root ($\Delta_x = 0$) of the quadratic Equation (10)

$$x_0 = \frac{1 - (\Delta + \beta_0)r}{2\beta_0 r} = \frac{dr}{d\tau} \tag{17}$$

to be equivalent to

$$\frac{dr}{d\tau} = \frac{1}{2\beta_0 r} - \frac{\Delta + \beta}{2\beta_0} \tag{18}$$

which, after denoting the radius by r_d, can be named the diffusion limit if

$$\frac{1}{r_d} - \Delta + \beta_0 > 0 \tag{19}$$

or

$$\frac{1}{r_d} > \Delta - \beta_0. \tag{20}$$

It implies that the critical value of the non-spherulitic or diffusional growth reads

$$r_d = \frac{1}{\Delta - \beta_0}. \tag{21}$$

The non-dimensional radius given by Equation (21) reflects an interplay between thermodynamic (Δ) and kinetic (β_0) parts at the onset of the diffusion-driven but non-spherulitic growth. Cleary, one can also expect such mode of growth if one puts $d(r) = 0$ in Equation (15), making then use of its equivalence with Equation (18) with zeroth condition applied to Equation (16).

The non-diffusive, thus, the spherulitic onset of the evolution, arises if one assumes a proportionality of $d(r)$ to r; thus, when providing a linearity thereof (but not a constancy with non-negative property), namely

$$d(r) \sim r \Rightarrow \; [(\Delta + \beta_0)r - 1]^2 + 4\beta_0 r(r+1) \sim r^2 \tag{22}$$

which implies that

$$(\Delta + \beta_0)^2 r^2 - 2(\Delta + \beta_0)r + 1 + 4\beta_0 r^2 + 4\beta_0 r \sim r^2 \tag{23}$$

and a simple limit of the form after postponing all in-r quadratic (counter-balancing) terms

$$2r(2\beta_0 - \Delta - \beta_0) + 1 \to 0 \tag{24}$$

can ultimately be taken, which results in

$$2r(\beta_0 - \Delta) \to -1 \tag{25}$$

and eventually leads to (denote the limit by r_{nd})

$$r_{nd} = \frac{1}{2(\Delta - \beta_0)}. \tag{26}$$

It is if $\beta_0 \approx 0.2$, cf. Figure 2, which is when the kinetics and thermodynamics work at the singularity-expressing onset of the spherulitic formation, with $\Delta \approx 0.2$, c.f. Equations (21) and (26).

Comparing Equation (26) with Equation (21), one immediately recognizes that a simple relation

$$r_d = 2 \cdot r_{nd} \tag{27}$$

prevails. Despite the physical fact that the factor two in proportionality relation (27) means an earlier onset (as compared to that non-spherulitic viz diffusional mode) of the spherulitic growth to manifest. The factor two can also be interpreted in graphical terms of Figure 1, namely, that in the diffusion (Mullins–Sekerka type) limit characteristic of r $\sim t^{1/2}$, the only one crystalline phase builds the object (Figure 1a), while in the case of the spherulitic growth, two concurrent phases (crystalline vs. amorphous) constitute the spherulite's body, cf. Figure 1b.

Thus, this way it has been shown that, at the onset of the growth under study and presumably around the kinetic-thermodynamic singularity $\beta_0 \sim \Delta$, cf. Equations (21) and (26), the spherulitic growth will prevail earlier in terms of the spherulite's nucleus value, cf. Equation (27), if the mass-convection and nonequilibrium (Goldenfeld type [3]) boundary conditions win over those of the Mullins–Sekerka kind. This feasible singularity limit at the onset of the growth, namely, $\beta_0 \sim \Delta$, suggests that, though the model is deterministic, its extension can fairly be envisioned towards applying prospectively a stochastic approach [15], wherein the corresponding fluctuations around the $\beta_0 \sim \Delta$ condition can show up in subtle or pronounced ways.

The overall spherulitic formation as viewed in the so-called entropy-production (e) terms [16] detected at the interface can be associated with the stochastic-fluctuational context. The scalar product $\vec{j}\,[c(R)] \circ \vec{n}_0$ put at the r.h.s. of Equation (1) involves the matter flux $\vec{j}\,[c(R)]$ of the external feeding field. The field is twofold, namely, either of mass-convective or of locally diffusional character. In the former, it does not include curvatures, whereas in the latter it receives them for granted, see Figure 1a, and it basically goes like $\vec{j}\,[c(R)] \circ \vec{n}_0 \sim 1/r$. For the mass-convective counterpart, one gets $\vec{j}\,[c(R)] \circ \vec{n}_0 \sim const$, when late times conditions readily apply. (In fact, in this time zone, the local curvatures of the Mullins–Sekerka mode also cease to grow, yielding ultimately a similar physical scenario.) The entropy production $e = \vec{j}\,[c(R)] \circ \vec{x}$ with $\vec{x}$ representing the (Onsager type) thermodynamic force enables, while based on the same reasoning, to ascertain that the involvement of (local) curvature term and the application of the Fick's law $\vec{x} \circ \vec{n}_0 \sim 1/r$ to e gives a bigger non-negative account to it based thoroughly on the Mullins–Sekerka [13,17] (crystal growth) mode than in the case of mass-convective, thus spherulitic mode, cf. Equation (3).

4. Conclusions

In this study, we have demonstrated, while based on the non-dimensional model (suitable for numerics), that upon mass-convection and nonequilibrium boundary criteria for the (poly)crystal's growth, such as the one of (bio)polymers addressing or that concerning biominerals (geophysical objects), realized in defects containing and condensed matter involving matrix, that a well-justifiable chance of spherulites' emergence prior to a pure diffusion-controlled crystal growth exists at the onset of the growing conditions. The argumentation line is based on the physical fact that, in spherulites (polycrystals) two phases may "synergistically" coexist, whereas in single crystals the only ordered crystalline phase has to be built in suitably, presumably at a higher energetic cost than in the former. As named by us, the unimodal crystalline Mullins–Sekerka type mode of growth, characteristic of local curvatures' presence, seems to be more entropy-productive in its emerging (structural) nature than the so-named bimodal or Goldenfeld type mode of growth in which the local curvatures do not play any crucial roles, and in which kinetics seems to win over thermodynamics. In turn, a liaison of amorphous and crystalline phases makes the system far better compromised to the thermodynamic-kinetic conditions it actually, and concurrently, follows.

The final conclusions presented are based on the peculiar evolution equation, having corroborated cooperatively mass-convective and nonequilibrium (boundary) conditions, that basically drive the growing system far from equilibrium. Interestingly, one may qualitatively predict that the entropy production [16] in such massive (poly)crystalline forms is more expressed in the unimodal non-compromised Mullins–Sekerka type mode. In this mode, a thermodynamic, close-to-equilibrium supersaturation factor prevails, unlike in its Goldenfeld-like bimodal nonequilibrium counterpart. This bimodality, i.e., a synergistic coexistence of crystalline and amorphous phases within a growing spherulite, rests on terms of internal stress–strain material conditions, and does not admit the local curvatures at the interface to prevail. The latter is not the case of any diffusion-controlled (unimodal) growth in which the so-called Mullins–Sekerka instability manifests readily. Finally, let us stress that it seems to us that the non-dimensionality of the proposed modeling suggests that the system does not depend upon experimental details, manifesting somehow a quasi-universal, that means, mainly the scaling addressing character of the performed modeling [18].

Author Contributions: Conceptualization, A.G. and J.S.; methodology, A.G.; software, J.S.; validation, A.G. and J.S.; formal analysis, A.G. and J.S.; investigation, A.G. and J.S.; resources, A.G. and J.S.; data curation, J.S.; writing—original draft preparation, A.G. and J.S.; writing—review and editing, A.G. and J.S.; visualization J.S. All authors have read and agreed to the published version of the manuscript.

Funding: This research was funded by BN-WTiICh-11/2022 of the Bydgoszcz University of Science and Technology.

Institutional Review Board Statement: Not applicable.

Informed Consent Statement: Not applicable.

Data Availability Statement: Not applicable.

Acknowledgments: On the anniversary of the 70th birthday of Jerzy Łuczka, we would like to thank him for many years of scientific cooperation and wish him good health and further success in his private and professional life.

Conflicts of Interest: The authors declare no conflict of interest.

References

1. Kittel, C. *Introduction to Solid State Physics*; Wiley: Hoboken, NJ, USA, 2004; ISBN 978-0-471-41526-8.
2. Crist, B.; Schultz, J.M. Polymer Spherulites: A Critical Review. *Prog. Polym. Sci.* **2016**, *56*, 1–63. [CrossRef]
3. Goldenfeld, N. Theory of Spherulitic Crystallization. *J. Cryst. Growth* **1987**, *84*, 601–608. [CrossRef]
4. Keith, H.D.; Padden, F.J. A Phenomenological Theory of Spherulitic Crystallization. *J. Appl. Phys.* **1963**, *34*, 2409–2421. [CrossRef]

5. Tseng, C.-H.; Tsai, P.-S. The Isothermal and Nonisothermal Crystallization Kinetics and Morphology of Solvent-Precipitated Nylon 66. *Polymers* **2022**, *14*, 442. [CrossRef] [PubMed]
6. Raimo, M. Growth of Spherulites: Foundation of the DSC Analysis of Solidification. *ChemTexts* **2015**, *1*, 13. [CrossRef]
7. Gránásy, L.; Pusztai, T.; Tegze, G.; Warren, J.A.; Douglas, J.F. Growth and Form of Spherulites. *Phys. Rev. E* **2005**, *72*, 011605. [CrossRef] [PubMed]
8. Gadomski, A.; Siodmiak, J.; Santamaria-Holek, I.; Rubi, J.M.; Ausloos, M. Kinetics of Growth Process Controlled by Mass-Convective Fluctuations and Finite-Size Curvature Effects. *Acta Phys. Pol. B* **2005**, *36*, 1537–1559.
9. Gadomski, A.; Grzywna, Z.J.; Łuczka, J. The Growing Processes in Diffusive and Convective Fields. *Chem. Eng. Sci.* **1993**, *48*, 3713–3721. [CrossRef]
10. Gadomski, A.; Łuczka, J. Convection-Driven Growth in Fluctuating Velocity Field. *Acta Phys. Pol. Ser. B* **1993**, *24*, 725–731.
11. Gadomski, A.; Łuczka, J. Some Remarks Concerning Spherulitic Growth. *Int. J. Quantum Chem.* **1994**, *52*, 301–308. [CrossRef]
12. Gadomski, A.; Łuczka, J. Spherulitic Growth of Polymer Systems in Convection Field under Non-Equilibrium Conditions. In *Far-From-Equilibrium Dynamics of Chemical Systems, Proceedings of the Third International Symposium*; World Scientific: Singapore, 1994; pp. 342–349.
13. Mullins, W.W.; Sekerka, R.F. Stability of a Planar Interface During Solidification of a Dilute Binary Alloy. *J. Appl. Phys.* **1964**, *35*, 444–451. [CrossRef]
14. Gadomski, A.; Siódmiak, J. A Novel Model of Protein Crystal Growth: Kinetic Limits, Length Scales and the Role of the Double Layer. *Croat. Chem. Acta* **2003**, *76*, 129–136.
15. Łuczka, J.; Niemiec, M.; Rudnicki, R. Kinetics of Growth Process Controlled by Convective Fluctuations. *Phys. Rev. E* **2002**, *65*, 051401. [CrossRef] [PubMed]
16. Reguera, D.; Rubí, J.M.; Vilar, J.M.G. The Mesoscopic Dynamics of Thermodynamic Systems. *J. Phys. Chem. B* **2005**, *109*, 21502–21515. [CrossRef] [PubMed]
17. Gadomski, A.; Siódmiak, J. A Kinetic Model of Protein Crystal Growth in Mass Convection Regime. *Cryst. Res. Technol.* **2002**, *37*, 281–291. [CrossRef]
18. Piorkowska, E.; Galeski, A.; Haudin, J.-M. Critical Assessment of Overall Crystallization Kinetics Theories and Predictions. *Prog. Polym. Sci.* **2006**, *31*, 549–575. [CrossRef]

entropy

Article

Information Processing Using Networks of Chemical Oscillators

Jerzy Gorecki

Institute of Physical Chemistry, Polish Academy of Sciences, Kasprzaka 44/52, 01-224 Warsaw, Poland; jgorecki@ichf.edu.pl

Abstract: I believe the computing potential of systems with chemical reactions has not yet been fully explored. The most common approach to chemical computing is based on implementation of logic gates. However, it does not seem practical because the lifetime of such gates is short, and communication between gates requires precise adjustment. The maximum computational efficiency of a chemical medium is achieved if the information is processed in parallel by different parts of it. In this paper, I review the idea of computing with coupled chemical oscillators and give arguments for the efficiency of such an approach. I discuss how to input information and how to read out the result of network computation. I describe the idea of top-down optimization of computing networks. As an example, I consider a small network of three coupled chemical oscillators designed to differentiate the white from the red points of the Japanese flag. My results are based on computer simulations with the standard two-variable Oregonator model of the oscillatory Belousov–Zhabotinsky reaction. An optimized network of three interacting oscillators can recognize the color of a randomly selected point with >98% accuracy. The presented ideas can be helpful for the experimental realization of fully functional chemical computing networks.

Keywords: chemical computing; network; oscillators; top-down design; Oregonator model; Japanese flag problem

Citation: Gorecki, J. Information Processing Using Networks of Chemical Oscillators. *Entropy* **2022**, *24*, 1054. https://doi.org/10.3390/e24081054

Academic Editor: Adam Gadomski

Received: 7 July 2022
Accepted: 26 July 2022
Published: 31 July 2022

1. Introduction

Chemical computers are everywhere because all living organisms use them for acquiring and transmitting information and for decision-making. Animals and humans, using chemical computers represented by their nervous systems and brains [1], are able to control complex life processes such as orientation in space, navigation in crowded environments, creating models of the environment they live in, developing self-awareness and even predicting the future. This demonstrates that Nature-made chemical computers can perform very complex computational tasks with low energy consumption.

The information-processing industry is based on semiconductor technology. The unprecedented success of this technology in machine information processing [2] is possible because semiconductor logic gates are highly reliable. They are characterized by a long time of error-free operation and can be downsized to the nanoscale [3]. The gates can be concatenated within a single integrated circuit and perform more complex information processing functions. These properties of semiconductor information-processing devices perfectly match the bottom-up design strategy, according to which more complex operations are represented by combinations of simpler tasks for which constructions of corresponding circuits have already been developed [4].

The concept of logic gates and binary information coding, so successfully used for information processing with semiconductor devices, has strongly influenced other fields of unconventional computation, including the use of chemicals [5]. There are many reports on chemical realization of logic gates or binary operations [6–14]. The fact that a chemical medium allows for implementation of all basic logic gates proves that a universal chemical computer can be constructed. In exceptional cases, molecular logic gates used as molecular probes offer an interesting alternative to standard techniques [14]. However, most chemical logic gates, especially those constructed with a reaction-diffusion medium [15], are neither small nor fast. For the Belousov–Zhabotinsky (BZ) reaction, the output signal of a gate appears a few seconds after the input is introduced [7,8,10]. For other chemical media, this time can be much longer. In the case of information coded in DNA molecules, it may take a few hours before the gate answer is obtained [16]. In typical experimental conditions, the time of reliable chemical gate operation is measured in hours, not years as for semiconductors. In my opinion, it would be difficult to make a chemical device in which many chemical gates are concatenated, communicate and demonstrate stable functionality for a long time.

I think the BZ reaction [17,18] is the most frequently used medium in experimental studies on chemical computation. The BZ reaction is a complex catalytic oxidation of an organic substrate (usually malonic acid) in an acidic environment [19,20]. Two stages of the BZ reaction can be visually identified. One is fast oxidation of the catalyst, and the other is slow reduction of the catalyst by an organic substrate. The solution color reflects concentrations of catalyst in the oxidized and reduced forms. Therefore, many types of nonlinear evolution of the medium as oscillations or spatio-temporal patterns can be easily observed. If the BZ reaction proceeds in a spatially distributed medium, then local excitation corresponding to a high concentration of the reaction activator can propagate in space in the form of a concentration pulse. This type of behavior resembles the propagation of nerve impulses in living organisms. As a result, the BZ reaction has attracted attention as a medium for experiments with neuron-like chemical computing [21–23].

Within the most popular approach to computing with BZ medium, it is assumed that information is transmitted by propagating pulses of the oxidized form of catalyst. For binary coding, the presence of a pulse represents the logic TRUE state, and the state with a low concentration of the catalyst in the oxidized form is the logic FALSE state [13,21]. If the ruthenium complex ($Ru(bpy)_3^{2+}$) is used as the reaction catalyst, then the BZ reaction becomes photosensitive [24,25] and can be externally controlled by illumination. Oscillations can be inhibited by light, which means that for the same initial concentrations of reagents, the medium oscillates in the dark, is excitable at a low light intensity, and shows a strongly attracting stable steady-state when illumination is strong. In a medium with the photosensitive BZ reaction, excitable channels through which signals can propagate can be formed by specific illumination of a spatially distributed medium. Using a suitable geometry of excitable and non-excitable channels, one can control the propagation of excitations and, for example, make a signal diode [26], a memory cell [27,28] or logic gates [13,15]. However, in typical applications, such gates are big (with an area of about 1 cm^2), and a single operation takes more than 10 s [20]. Therefore, the bottom-up approach from gates to complex information processing tasks does not look promising if binary information coding is used with BZ medium.

Literature reports demonstrate that a chemical medium can be efficiently applied for specific computing tasks. Parallel processing of information by the medium as a whole is a common feature of efficient implementations. The classic example is the Adleman experiment proving that the Hamiltonian path problem can be solved with DNA molecules [29,30]. Another example is the so-called prairie-fire algorithm for verifying if there is a path linking two randomly selected points in a labyrinth. This problem can be solved by a labyrinth formed from an excitable medium where stable pulses of excitation can propagate [31,32]. If there is a path linking two points, an excitation generated at one of the points will then appear at the other, and the time difference between excitation and detection can be used to estimate the shortest path linking these points. Yet another famous computing application of a chemical medium working in parallel is the image processing of black and white photos performed using a photosensitive variant of the BZ reaction proceeding in a uniform, spatially distributed system [33,34]. In such a medium, image processing is the consequence of a non-homogeneous initial state generated by initial illumination with intensity proportional to the grayscale of pixels of the processed image. In all methods mentioned above, the output information is coded in the time evolution of the computing medium.

However, the number of man-written parallel algorithms that can be efficiently executed using a chemical medium of parallel algorithms for chemical computers is very limited. A top-down design strategy offers a promising approach for finding new ones. The strategy can be summarized as follows. In the beginning, we select a problem we want to solve and the computing medium that is supposed to do it. Next, we define how the input information is introduced and how the output is extracted from observing medium evolution. The top-down approach can be applied if the properties of the medium—and thus of the medium evolution—can be controlled by a number of adjustable parameters. Within this strategy, we are supposed to find the values of parameters for which the medium answer (the output) gives the most accurate solution to the considered problem. To perform such optimization, we need a number of examples (the training dataset) that can be used to verify the accuracy of computation performed by the medium.

Networks of interacting chemical oscillators seem to be an interesting candidate for a chemical computing medium. Networks of BZ oscillators can be assembled using droplets containing reagents stabilized by lipids dissolved in the surrounding oil phase [35–37]. The idea of information processing with networks of interacting chemical reactors was formulated in [38]. In such an approach, a node is defined by a set of reactions sharing the specific reagents. Interaction between nodes consists of reactions providing the exchange of reagents between nodes. In [38–41], the authors focused on nodes that show excitable or bistable behavior; thus, the concentration of reagents in a single node can evolve towards one of two values that can be interpreted as corresponding to binary logical values [6]. It has been demonstrated that such nodes can perform logic gate operations. The idea of computing with oscillator networks generalizes the approach described above. First, the dynamics of a node are more complex. Second, the node answer is not a stationary state but reflects the character of evolution observed within a finite time interval. As the system is continuously evolving, the time of observation is an important parameter.

Several theoretical studies demonstrate the computing potential of oscillator networks [42–44]. It has been shown that oscillator networks could be optimized to perform classification tasks [45,46] and process information with the best possible use of the chemical medium. In this paper, I am concerned with the previously reported determination of color for a randomly selected point on the Japanese flag [47]. I introduce a few new elements to computing oscillator networks, including the comparison between two Oregonator models for node evolution that exhibit the different character of oscillations and generalization of the node–node interaction model with coupling parameters individually adjusted for pairs of nodes. Moreover, a new concept of extracting the output information from the network is proposed. In all previous papers on the subject, the number of activator maxima observed on a selected network node was regarded as the network answer. Here the total amount [48] of activator or inhibitor observed on the output oscillator is regarded as the network output. The results presented below demonstrate that such an approach is equally useful and leads to similar accuracy in determining the color of a point on the flag based on its coordinates as the number of activator maxima.

The paper is organized as follows. The information on the computational problem I am concerned with, the mathematical model of the time evolution of a node and of the network, and the optimization procedure are described in Section 2. Section 3 contains obtained results and their discussion. The final section presents verification of obtained results and presents suggestions for future studies.

2. Information Processing with Oscillator Networks

In this section, I present general information on the types of problems that can be solved with oscillator networks. Moreover, I discuss the network structure and a chemical model used to simulate its evolution.

2.1. Classification Type Problems

In a number of recent papers, it has been shown that networks of interacting chemical oscillators can be trained to perform classification tasks with reasonable accuracy [45,46,49]. Let us consider a problem A defined by a set of records $D_A = \{r_n, n = 1, N\}$. Each record $r_n = (p_n^1, p_n^2, \ldots, p_n^k, s_n)$ is in the form of a $(k+1)$ tuple, where the first k elements are predictors represented by real numbers, and the last element (s_n) is the record type, and it is represented by an integer. It is assumed that the number of possible predictor values is limited. Let D_A denote a database of records related to problem A. The classifier of D_A is supposed to return the correct data type if the predictor values are used as its input.

There are classification problems for which D_A is finite. For example the AND logic gate is equivalent to classification of the database: $D_A = \{\{0,0,0\},\{1,0,0\},\{0,1,0\},\{1,1,1\}\}$. In this paper, I consider a geometrically inspired problem of determining the color of a randomly selected point located on the Japanese flag (cf. Figure 1). The red disk (sun) is centrally located in a white square (here represented by the Cartesian product $[-0.5, 0.5] \times [-0.5, 0.5]$). Let us notice that the location of the Japanese flag differs from the one considered in our previous paper on the problem [47] where it was $[0.0, 1.0] \times [0.0, 1.0]$. This has been done intentionally to see if object location can influence classifier accuracy. The disk radius is $r = 1/\sqrt{(2\pi)}$; thus, the areas of the sun and the white region are equal. Records of the considered problem have the form: (x, y, s), where $(x, y) \in [-0.5, 0.5] \times [-0.5, 0.5]$ are the point coordinates, and the record type $s \in \{0, 1\}$ tells if the point (x, y) is located in the red $s = 1$ or in the white region $s = 0$. A network that gives a random answer or a network that always gives the same answer ("the point is red" or "the point is white") to all inputs solves the problem with 50% accuracy (or with a 50% chance of obtaining the wrong answer). I postulate that the Japanese flag problem can be solved with much higher accuracy by a network of chemical oscillators.

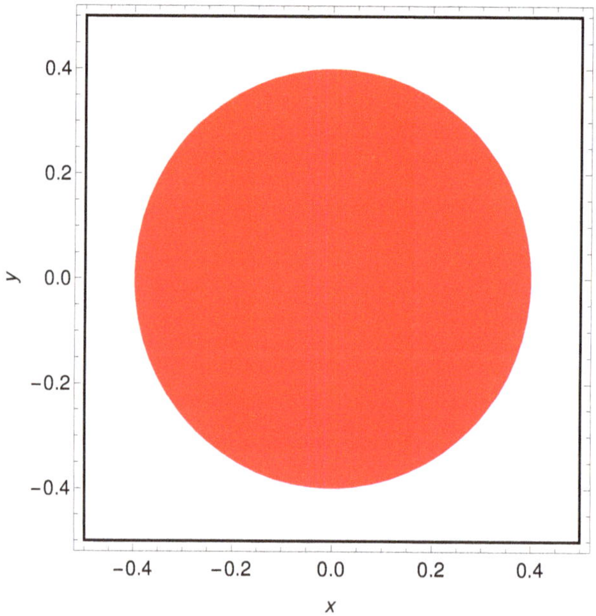

Figure 1. The geometrically inspired problem of determining the color of a randomly selected point located on the Japanese flag formed by the central red disk and the surrounding white area. The flag is represented by the Cartesian product $[-0.5, 0.5] \times [-0.5, 0.5]$), and the disk radius is $r = 1/\sqrt{(2\pi)}$; thus, the areas of the sun and the white region are equal.

2.2. The Node Model

Before applying the top-down network optimization strategy, we should select the medium that is supposed to perform the classification. Here I use the two-variable Oregonator model [50] of the BZ reaction to describe the time evolution of an individual oscillator. The model equations are:

$$\frac{du}{dt} = \frac{1}{\varepsilon}(u(t) - u(t)^2 - (fv(t) + \phi(t))\frac{u(t) - q}{u(t) + q}) - \alpha u(t) \quad (1)$$

$$\frac{dv}{dt} = u(t) - v(t) \quad (2)$$

where u and v, respectively, represent concentrations of the reaction activator U corresponding to $HBrO_2$ and inhibitor V that in the two-variable Oregonator model is the oxidized form of the catalyst. The time evolution of a medium where the BZ reaction proceeds is determined by the values of parameters: f, q and ε. The parameter ε sets up the time scale ratio between variables u and v, q is the scaling constant, and f is the stoichiometric coefficient. The time-dependent function $\phi(t)$ that appears in Equation (1) is related to medium illumination. The Oregonator model is computationally simple, and it allows for performing complex evolutionary optimization involving the massive number of evaluations of network evolution needed for evolutionary optimization. It takes into account the effect of the combined excitation of one node by a few neighbors. Despite its simplicity, the Oregonator model provides a better-than-qualitative description of many phenomena related to the BZ reaction. It correctly describes the oscillation period as a function of reagent concentration and also can be used to simulate nontrivial phenomena such as the migration of a spiral in an electric field [51] or reaction of a propagating pulse to time-dependent illumination [52]. Of course, a model with a larger number of variables gives a more realistic description of the BZ reaction but, on the other hand, requires a more precise model of interactions between oscillators.

The last term in Equation (1) describes the activator decay, and it does not appear in the standard form of the Oregonator model. This term can be related to a reaction:

$$U + D \rightarrow products \quad (3)$$

where D describes the other reagents of the process that occur with the rate α. As I discuss later, the existence of this process is equivalent to the presence of a "sink" node in the network that adsorbs the activator molecules.

The reported simulations of network evolution have been performed for two different sets of Oregonator model parameters. The time evolution in one of the considered networks is described by Model I: $\varepsilon = 0.2$, $q = 0.0002$ and $f = 1.1$, which was used in the previous study on the color of a point on the Japanese flag [47]. For this set, the period of oscillations is $\sim$10.7 time unit if $\phi(t) = 0$. The optimization of other networks was done for the Oregonator Model II defined by: $\varepsilon = 0.3$, $q = 0.002$ and $f = 1.1$. The period of oscillations is $\sim$8 time unit for $\phi(t) = 0$. The character of activator and inhibitor oscillations for the above-mentioned models and $\alpha = 0.7$ or $\alpha = 0.5$ is illustrated in Figure 2. For both sets of parameters, the system converges to a stable stationary state for $\phi \sim 0.2$ [25,53].

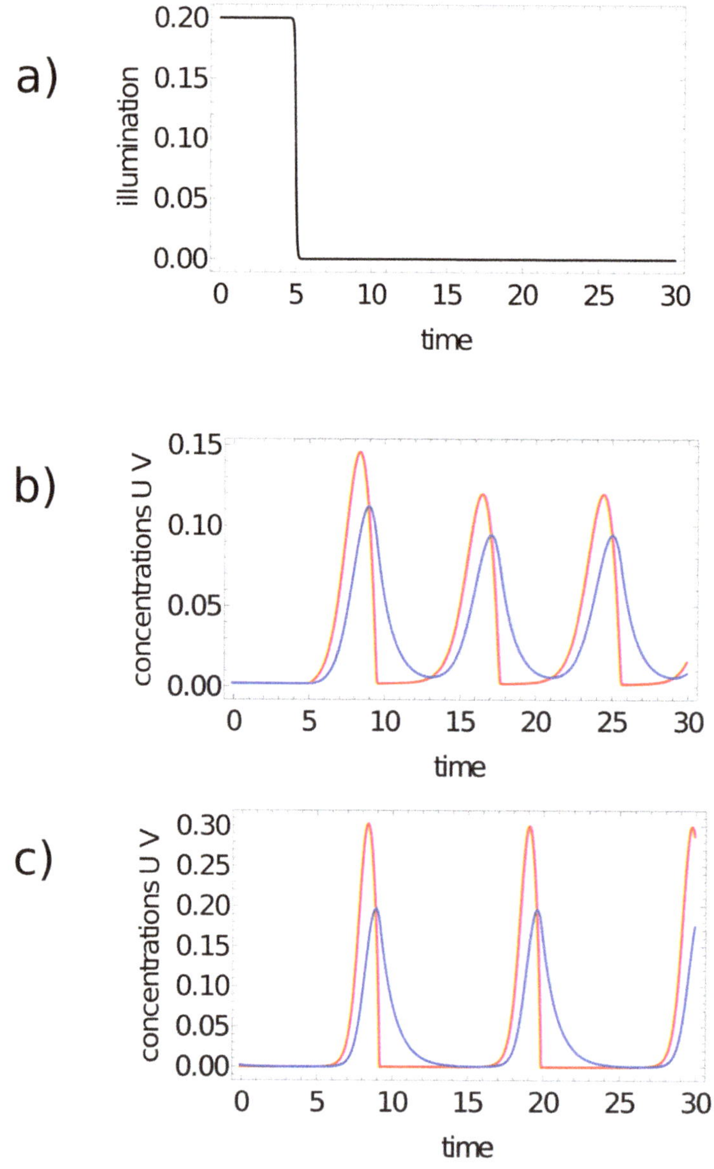

Figure 2. (**a**) Time-dependent illumination $\phi(t) = (1.001 + \tanh(-10 * (t - t_{illum})))/10$ for $t_{illum} = 5$. (**b**,**c**) The character of oscillations for the 2-variable Oregonator models used in simulations: (**b**) Model II: $f = 1.1$, $q = 0.002$, $\epsilon = 0.3$; (**c**) Model I: $f = 1.1$, $q = 0.0002$, $\epsilon = 0.2$. Red and blue curves represent concentrations of activator (u) and inhibitor (v), respectively. The values of α are 0.5 (**b**) and 0.7 (**c**).

The value of ϕ can be interpreted as the light intensity in the Ru-catalyzed BZ reaction, and it can be used as an external factor to suppress or restore oscillations. For the control of computing networks discussed below, I considered $\phi(t)$ in the form:

$$\phi(t) = W \cdot (1.0 + \eta + \tanh(-\xi \cdot (t - t_{illum})) \quad (4)$$

Such a definition of $\phi(t)$ involves a few parameters: W and η determine the limiting values of illumination at $t \to \pm\infty$. The value of illumination time t_{illum} defines the moment of time when the most rapid changes in illumination occur, and ξ represents the rate at which the transition occurs. The minus sign in the argument of the tanh() function implies the transition from steady state (high illumination) towards oscillations (low illumination). In the presented simulations, I used fixed values: $W = 0.1$, $\eta = 0.001$ and $\xi = 10$. For such parameters, the value of $\phi(t)$ is high (~0.2) in the time interval $[0, t_{illum} - \delta]$ ($\delta = 0.1$) (cf. Figure 2a). Oregonator models with parameters given above predict stable steady state. For long times ($t > t_{illum} + \delta$), the value of $\phi_j(t)$ approaches 0.0001, which corresponds to oscillations.

The use of illumination time (or, in general, the inhibition time for oscillations inside a node) t_{illum} to control oscillators is inspired by experiments in which oscillations of individual BZ droplets were regulated by blue light [54]. In these experiments, two illumination levels were used: a low one, for which the droplet was oscillating; and a high one, for which oscillations were inhibited. The transitions between steady state and oscillations predicted by the two-variable Oregonator model were in qualitative agreement with experiments.

2.3. The Model of a Network

An example illustrating the idea of a considered computing oscillator network is shown in Figure 3. The network is formed by three coupled chemical oscillators (nodes) graphically represented by circles. The nodes have different functions in networks. The upper one (#1) is a normal node whose illumination time is fixed. Two bottom nodes (#2) and (#3) are inputs of x and y coordinates, respectively. The arrows interlinking the oscillators represent reactions that exchange the activators between nodes. The arrows directed away mark the activator decay.

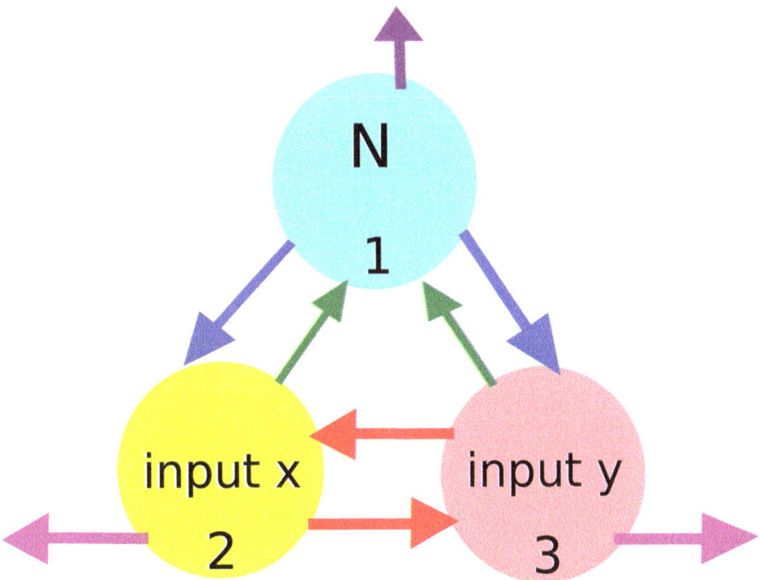

Figure 3. The idea of a computing oscillator network. Circles represent network nodes that are chemical oscillators. The nodes have different characteristics. The upper one (#1) is a normal one, and its illumination function is fixed. The bottom nodes (#2) and (#3) are inputs of x and y coordinates, respectively. The arrows interlinking oscillators represent reactions that exchange the activators between nodes. The arrows directed away mark activator decay (reaction 3).

If the jth oscillator is a normal one, then the value of $t_{illum}(j)$ that appears in the definition of $\phi(t)$ (Equation (4)) does not depend on input values. Normal nodes are supposed to moderate the interactions within the network. The set of their illuminations can be regarded as a program executed by the network. If an oscillator is considered the input of a predictor p^i. then the value of $t_{illum}(j)$ is functionally related to p^i. For the analysis presented below, it is assumed that this function is defined by two parameters, t_{start} and t_{end}, and has the form:

$$t_{illum}(j) = t_{start} + (t_{end} - t_{start}) \cdot p^i \tag{5}$$

Keeping in mind the symmetry of the Japanese flag problem, the values of t_{start} and t_{end} are the same for all predictors corresponding to x- and y-coordinates. This means that for record (x_n, y_n, s_n):

$$t_{illum}(2) = t_{start} + (t_{end} - t_{start}) \cdot x_n \tag{6}$$

and

$$t_{illum}(3) = t_{start} + (t_{end} - t_{start}) \cdot y_n \tag{7}$$

In a general case the values of t_{start} and t_{end} can be different for different predictors. The above form of relationship between the predictor value and illumination is simple. More complex functions can be used in specific cases and produce a network with higher accuracy. However, if the relationship between the input and the illumination time is represented by a complex function, then one can argue that part of the computation is done, not by the network, but by the specific form of this relationship.

Coupling between oscillators, indicated by arrows in Figure 3, is achieved by reactions that extend the original Oregonator model. I assume that the coupling is of the activatory type and occurs via the exchange of reactor activators between oscillators [49]. Let U_i denote the activator of the ith node. The interactions between the nodes #k and #j appear as the results of reactions involving the activators U_k and U_j:

$$U_j + B_{j,k} \rightarrow U_k + C_k \tag{8}$$

$$U_k + B_{k,j} \rightarrow U_j + C_j \tag{9}$$

with the reaction rate constants $k_{B,j,k}$ and $k_{B,k,j}$.

The changes in concentrations of U_k and U_j as the result of reactions (8) and (9) are described by:

$$\frac{du_j}{dt} = -k_{B,j,k} b_{j,k} u_j \tag{10}$$

$$\frac{du_k}{dt} = -k_{B,k,j} b_{k,j} u_k \tag{11}$$

and the changes in concentration of U_j as the result of reaction (3) by:

$$\frac{du_j}{dt} = -k_D d_j u_j \tag{12}$$

In Equations (10)–(12), $b_{j,k}$, $b_{k,j}$ and d_j denote concentrations of $B_{j,k}$, $B_{k,j}$ and D_j, respectively. We assume that these concentrations are high with respect to the concentrations of the activators involved, are the same for all oscillators, and remain almost constant during the network evolution. Therefore, they are not included in the model of network evolution. Let us introduce symbols α_j and $\beta_{j,k}$, defined as: $\alpha_j = k_D d_j$ and $\beta_{j,k} = k_{B,j,k} b_{j,k}$. Keeping in mind that values of α_j and $\beta_{j,k}$ can be modified by concentrations of D_j and $B_{j,k}$, we can treat them as free parameters that can be easily adjusted. The values of α_j and $\beta_{j,k}$ are included in the optimization procedure. Let us also notice that all information on the network geometry is included in the values of $\beta_{j,k}$, because for non-coupled nodes $\beta_{j,k} = \beta_{k,j} = 0$.

On the basis of the above assumptions, we can write equations describing the time evolution of the network:

$$\frac{du_j}{dt} = \frac{1}{\varepsilon}(u_j - u_j^2 - (fv_j + \phi_j(t))\frac{u_j - q}{u_j + q}) - (\alpha_j + \sum_{i=1,m} \beta_{j,i})u_j + \sum_{i=1,m} \beta_{i,j}u_i \quad (13)$$

$$\frac{dv_j}{dt} = u_j - v_j \quad (14)$$

where i,j represent the ith and jth oscillator, respectively, and m is the number of oscillators in the network. This set of equations is solved numerically with a fourth-order RK algorithm [55] and time step of $dt = 0.0005$.

I assume that the output information can be extracted from observation of the network evolution during the time interval $Z = [0, t_{max}]$. This assumption is essential. In many previous studies on chemical computing, it was assumed that the system reaches a specific steady-state that represents the answer. In the present approach, the output information can be read from the time evolution of a selected oscillator observed in a finite time interval, and what happens later is irrelevant to computation.

Let us notice that Equation (5) has a physical meaning for any value of $t_{illum}(j)$. If $t_{illum}(i) < 0$, then $\phi_i(t)$ is small and the oscillator #i is active during the whole observation interval. When $t_{illum}(k) > t_{max}$, then $\phi_k(t)$ is large and the oscillator #k is inhibited within Z and does not oscillate.

In previous papers on classification with networks of interacting chemical oscillators, the number of activator maxima observed within the time interval $Z - \{0, t_{max}\}$ on a selected node (say #j) was considered as the output [47,49]. However, there is a question: What is the maximum of activator? I assume that $u_j(t)$ should have a maximum in a strict mathematical sense. This means that if $u_j(t)$ has a maximum at t_0, then there exists $\nu > 0$ such that $u_j(t) < u_j(t_0)$ for all $t \in [t_0 - \nu, t_0 + \nu] \subset Z$. Therefore, if $u_j(t)$ is growing at the end of Z, then $u_j(t_{max})$ is not regarded as a maximum. Moreover, the value of $u_j(t_0)$ should be larger than the activator concentration in the part of the oscillation cycle when the catalyst is in its reduced form. Here, I assume that the threshold value is 0.05. To illustrate maxima counting, let us consider the network with parameters given in the first row of Table 1. Figures 4 and 5 show the time evolution of activator (red curves) and inhibitor (blue curves) at all nodes for two selected points of the flag. The green line marks the threshold for the activator maximum (0.05). The results in Figure 4 illustrate the time evolution of activator and inhibitor on all nodes when the coordinates of the input point are: $(-0.25, 0.25)$. The point characterized by such coordinates is located in the red area of the flag. For this input, there are two maxima of the activator at all network nodes. It can also be seen that on all nodes, the concentration of the activator is larger than 0.05 at the final stage of the evolution. However, the function $u(t)$ is still growing at the end of Z, so the third maximum of $u(t)$ is not observed. Figure 5 illustrates the time evolution of activator and inhibition on all nodes if the coordinates of the input point are: $(0.29, -0.29)$. This point is located in the white region of the flag. Now, we observe three maxima of the activator at nodes #1 and #3, and two maxima at node #2. Therefore, if we consider the number of maxima on a selected oscillator as the network answer, then by selecting nodes #1 or #3, one can distinguish which of two points $\{(-0.25, 0.25), (0.29, -0.29)\}$ is white and which is red. Such classification is not possible if node #2 is considered the output. Selection of the output node can be made on the basis of correct answers for the records included in the training dataset.

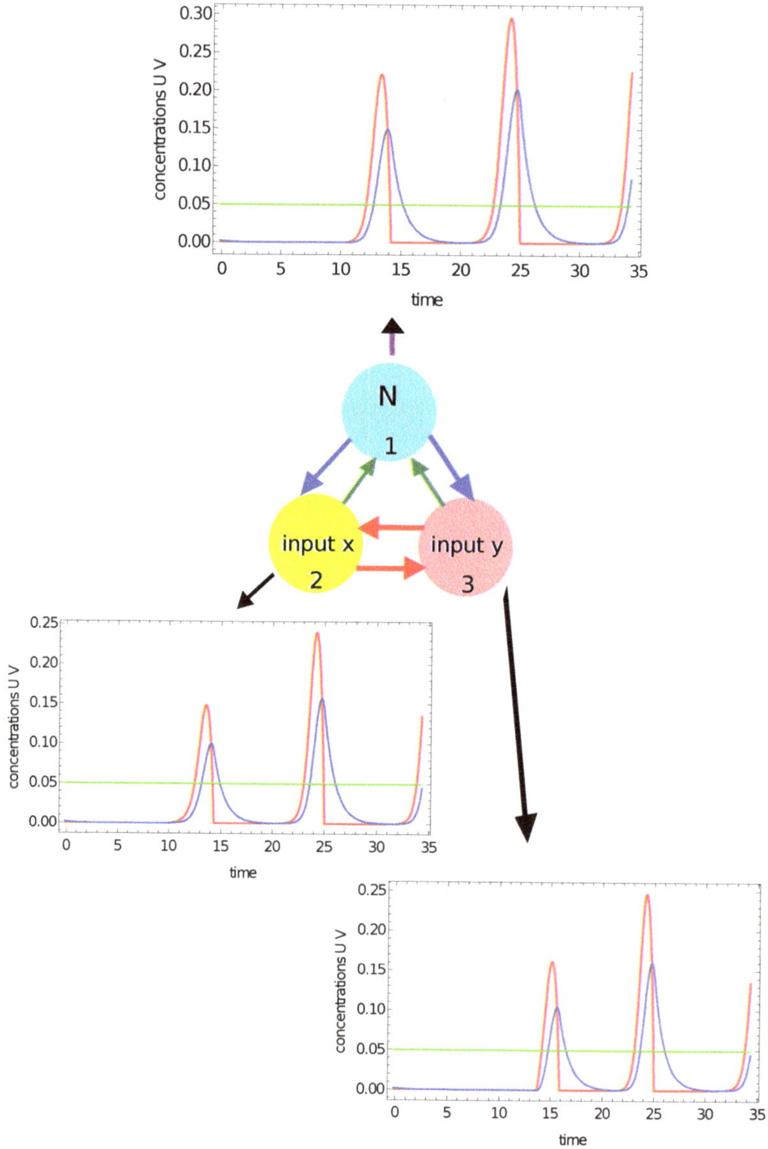

Figure 4. The time evolution of activator (the red curves) and inhibitor (the blue curves) observed on all nodes of the network defined by the parameters listed in the first line of Table 1. The coordinates of the input point are: $(-0.25, 0.25)$. The green line marks the threshold for the activator maximum. There are 2 maxima at all nodes in the network.

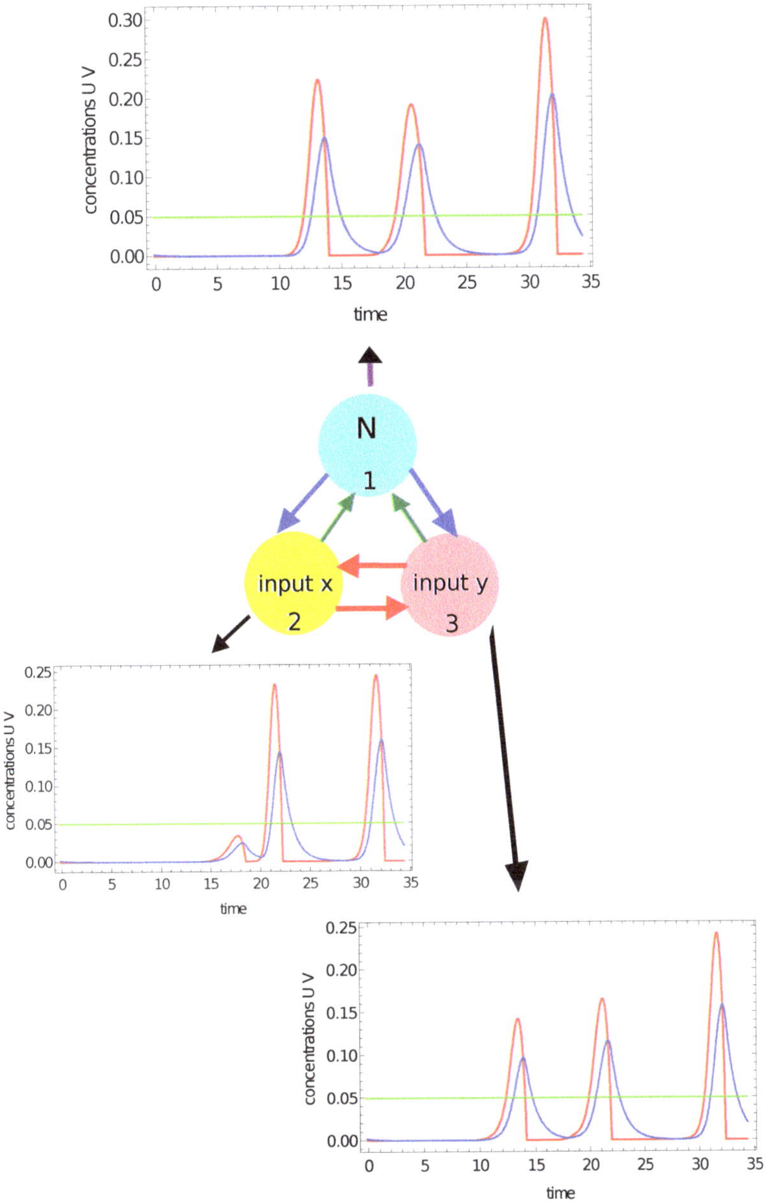

Figure 5. The time evolution of activator (the red curves) and inhibitor (the blue curves) observed on all nodes of the network defined by the parameters listed in the first line of Table 1. The coordinates of the input point are: $(-0.29, 0.29)$. The green line marks the threshold for the activator maximum. There are 3 maxima of $u(t)$ on nodes #1 and #3 and two maxima of $u(t)$ on node #2 within the observation time $[0, t_{max}]$. The green line marks the threshold for the activator maximum.

Figure 6 shows another example of classification with a number of activator maxima recorded on the selected node. Here, we consider the time evolution of the activator (the red curve) and the inhibitor (the blue curve) on node #1 of the network defined by the parameters listed in the second line of Table 1. Figure 6a illustrates the time evolution of concentrations for the point outside the red area $(-0.25, 0.28)$, and Figure 6b for the point inside it $(-0.39, -0.43)$. In the first case, no maximum of activator concentration is observed for times in $Z - \{0, t_{max}\}$; in the second case, we have a single maximum. Therefore, by observing the number of maxima on node #1, one can distinguish a red point from a white point, and a binary answer (0 or 1) is given. A classifying network with such properties can be regarded as highly optimized because the answer is obtained within a time similar to a single period of oscillation.

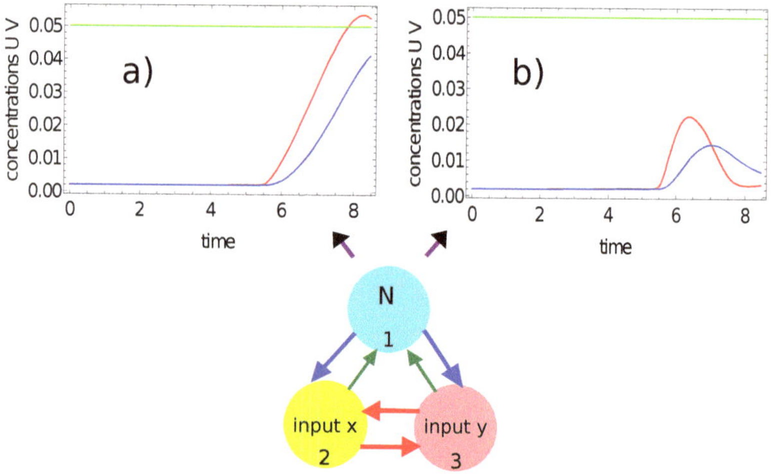

Figure 6. The time evolution of the activator (the red curve) and the inhibitor (the blue curve) on node #1 of the network defined by the parameters listed in the second line of Table 1: (**a**) $u_1(t)$ and $v_1(t)$ for the point inside the red area $(-0.25, 0.28)$; (**b**) $u_1(t)$ and $v_1(t)$ for the point outside the red area $(-0.39, -0.43)$.

Figures 7 and 8 illustrate another method for extracting the output information from the network evolution. This was inspired by [48], in which the output of an information-processing chemical BZ oscillator was related to the concentration of a selected reagent integrated over a specific time interval. Here we use the concentration of activator $u(t)$ (red, Figure 7) and the concentration of inhibitor $v(t)$ (blue, Figure 8) integrated over the observation time $[0, t_{max}]$ as the network output. In these figures, the shaded areas below the function represent the integrals $J_u = \int_0^{t_{max}} u_1(t)dt$ and $J_v = \int_0^{t_{max}} v_3(t)dt$, respectively. The integral value J is a real number, whereas an integer output is expected. In order to get such an output, I applied the following transformation:

$$output = floor(40 \cdot J) \tag{15}$$

where J stands for J_u or J_v. For the integrated concentration of the activator (Figure 7), we obtained $J_u = 0.101$ for the considered point inside the red area and $J_u = 0.073$ for the point outside it. Therefore, the network answers were 4 and 2 for these points, respectively. The method applied to the integral of the inhibitor (Figure 8) produced the outputs $J_v = 0.203$ for the considered point inside the red area and $J_u = 0.181$ for the point outside it. In this case, the network outputs were 8 and 7, respectively. Therefore, the integrals of the activator and inhibitor can also be used to classify points of different colors on the Japanese flag.

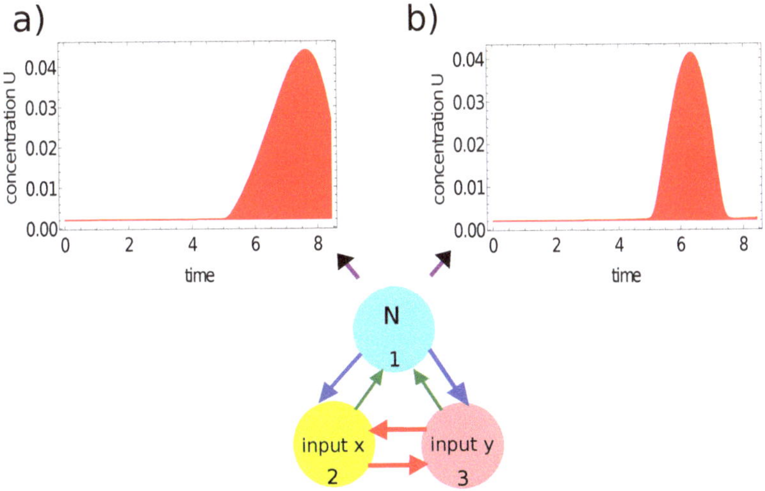

Figure 7. The time evolution of the activator at node #1 of the network defined by the parameters listed in the third line of Table 1: (**a**) $u_1(t)$ for the point $(-0.25, 0.28)$ located inside the red area; (**b**) $u_1(t)$ for the point $(-0.39, -0.43)$ located outside the red area. The red shaded area below the function represents the integral of $J_u = \int_0^{t_{max}} u_1(t)dt$, considered as the network output.

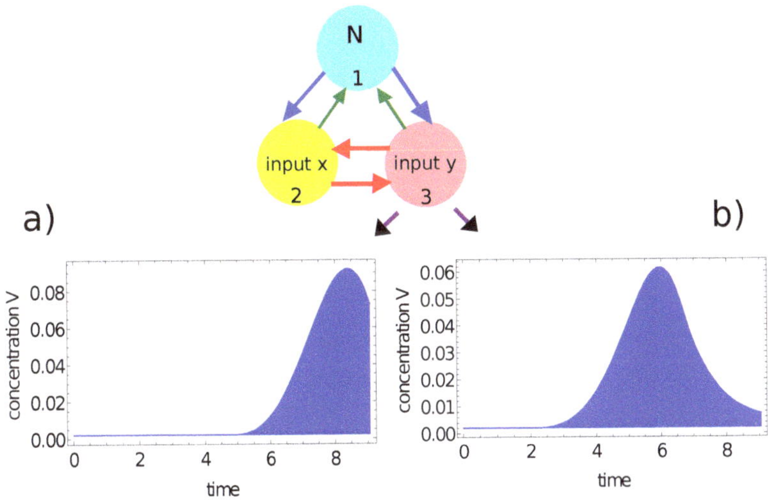

Figure 8. The time evolution of the inhibitor at node #3 of the network defined by the parameters listed in the fourth line of Table 1: (**a**) $v_3(t)$ for the point $(-0.25, 0.28)$ located inside the red area; (**b**) $v_3(t)$ for the point $(-0.39, -0.43)$ located outside the red area. The blue shaded area below the function represents the integral $J_v = \int_0^{t_{max}} v_3(t)dt$, considered as the network output.

Table 1. The parameters of networks that give the best correlations between the time evolution of the output oscillator and the point color.

Oregonator	Method	t_{max}	t_{start}	t_{end}	$t_{illum}(1)$	α_1	$\alpha_2 = \alpha_3$	$\beta_{1,2} = \beta_{1,3}$	$\beta_{2,1} = \beta_{3,1}$	$\beta_{2,3} = \beta_{3,2}$
Model I	activator maxima	34.4	11.6	19.9	10.1	0.87	0.72	0.16	0.43	0.29
Model II	activator maxima	8.45	3.77	8.03	5.42	0.96	0.46	0.53	0.38	0.42
Model II	u-integral	8.43	3.77	7.41	5.00	0.65	0.50	0.83	0.26	0.29
Model II	v-integral	9.06	3.77	7.56	5.71	0.75	0.44	0.60	0.29	0.33

2.4. Top-Down Design of Computing Networks

To define an information-processing chemical oscillator network, we have to specify many parameters:

- The observation time t_{max};
- All parameters for a model of chemical oscillations inside a node; for the Oregonator model, they are ε, q and f;
- Parameters t_{start} and t_{end} that translate an input value into the illumination of an input oscillator (cf. Equation (5));
- The rates for reactions responsible for interactions between oscillators (α_j, $\beta_{j,i}$);
- Location of input and normal oscillators;
- Finally, the illumination times for all normal oscillators $t_{illum}(i)$.

Obviously, a network with randomly selected parameters has a small chance of working as a good classifier. All parameters listed above should be optimized for executing the required function. I do not know any algorithm that allows for a straightforward design of the optimum oscillator network for a given problem. Still, we can apply the idea of supervised training for parameter optimization. Training means that we need a teacher, and in our optimization, it is a specific database T_A that contains a sufficient number of records related to the considered problem A [56].

The network should include the output oscillator whose time evolution is transformed into the network answer. Here I consider two types of answers. One is the number of activator maxima observed in the time interval $[0, t_{max}]$. Another is the integral of the activator or inhibitor concentration. If the network parameters are known, the output oscillator can be located.

In order to find which oscillator should be used as the output, one can calculate the mutual information $I(S;O_j)$ [57] between the discrete random variable S of record types in the training dataset T_A ($S = \{s_n, n = 1, N\}$) and the discrete random variable O_j representing the output of the jth oscillator in the network when the predictors of nth database record are used as the network input ($O_j = \{o_j(n), n = 1, N\}$). The mutual information $I(S;O_j)$ can be calculated as:

$$I(S;O_j) = H(S) + H(O_j) - H(S,O_j) \tag{16}$$

where $H()$ is the Shannon information entropy [58], and the discrete random variable $(S,O_j) = \{(s_n, o_j(n)), n = 1, N\}$. The oscillator #$i$ for which the mutual information between S and O_i is maximal is used as the network output. The mutual information calculated for the output oscillator is considered the measure of network fitness:

$$Fitness = max_{j \in \{1,2,3\}}\ I(S;O_j) \tag{17}$$

It can be expected that in the majority of cases, optimization based on mutual information leads to a classifier with the highest accuracy [59].

The fitness function based on mutual information does not require specifying how the network evolution translates into the network output. However, if we know how to link the output $o_j(n)$ with the record type, then we can calculate the accuracy on node #j for training dataset E_j as the ratio between correctly classified cases and all cases of T_A. For example, we can relate a specific number of activator maxima to the color of a point based on the majority of cases obtained for the training dataset. Therefore, we can alternatively define $Fitness$ as:

$$Fitness = max_{j \in \{1,2,3\}} E_j \tag{18}$$

Both formulae (17) and (18) allow us to locate the output oscillator in the network.

It has been demonstrated that evolutionary optimization [60] oriented towards obtaining the best classifier for a representative training dataset of the problem can lead to a computing network that performs the anticipated task with reasonable accuracy [45,46,49]. In this approach, we represent the parameters as a code that undergoes recombination between the fittest members and mutations of an offspring. For recombination, two networks are selected, and their parameters are randomly separated into two parts. Next, an offspring is generated by combining one part of the first network with the other part of the second one. At this step, the function of an oscillator (input, normal) and illumination times of normal oscillators are copied to the offspring. For the next step, mutation of all parameters of the newborn offspring is considered. The probability of mutation rate is 0.5 per step, and the change in parameter value does not exceed 20%. Each generation of networks consists of the same number of elements. It includes a few fittest networks from the previous generation that are copied without changes in parameters. The remaining members of the next generation are offspring created by recombination and mutation operations applied to oscillators from the top 50% of networks from the previous population.

However, optimization of all parameters as mentioned above represents a computational problem of very high complexity. Before starting optimization, we introduce a number of simplifications:

(1) My attention is restricted to classifiers formed by $m = 3$ oscillators;
(2) There have to be input oscillators for each coordinate in the network and a normal oscillator. Keeping in mind the symmetry of the considered network, we can assume that node #1 is the normal oscillator and nodes #2 and #3 are the inputs of x- and y-coordinates, respectively;
(3) The system symmetry reduces the number of parameters in the networks because: $\alpha_2 = \alpha_3$, $\beta_{1,2} = \beta_{1,3}$, $\beta_{2,1} = \beta_{3,1}$ and $\beta_{2,3} = \beta_{3,2}$.

After all these simplifications, the network is fully defined by the parameters listed in Table 1.

3. What Is the Color of a Point on the Japanese Flag? (As Seen by the Networks)

Here I present the results of network optimization. The training dataset is composed of 1000 records; 501 represent points inside the sun area and 499 are points outside it. The location of points can be seen, for example, in Figure 9. The applied evolutionary algorithm is a standard one, and it has been described in [45,46,49]. Optimization starts with 80 networks with randomly selected parameters. All networks in the initial population are evolved for maximum fitness. The fittest 10% (8) of the networks are copied to the next generation. The population of the next generation is completed to 80 networks by classifiers obtained by recombination and mutation operations applied to networks randomly selected from the upper 50% of the fittest. Next, the evolution step is repeated on the new population. The optimization procedure is executed for 1000 steps.

Figures 9–12 illustrate the answer of the fittest networks obtained for two Oregonator models and three different methods of interpreting the network output. The parameters of the discussed networks are given in Table 1. In all cases, the network output is based on the time evolution of a selected node in the network. For the first and second networks (Figures 9 and 10), this is the number of activator maxima observed on a selected node in the time interval $[0, t_{max}]$. For the third and fourth network, this is the integrated concentration of activator on a selected node (Figure 11) and the integrated concentration of inhibitor on a selected node (Figure 12). In all networks, the classification rule is derived after network optimization.

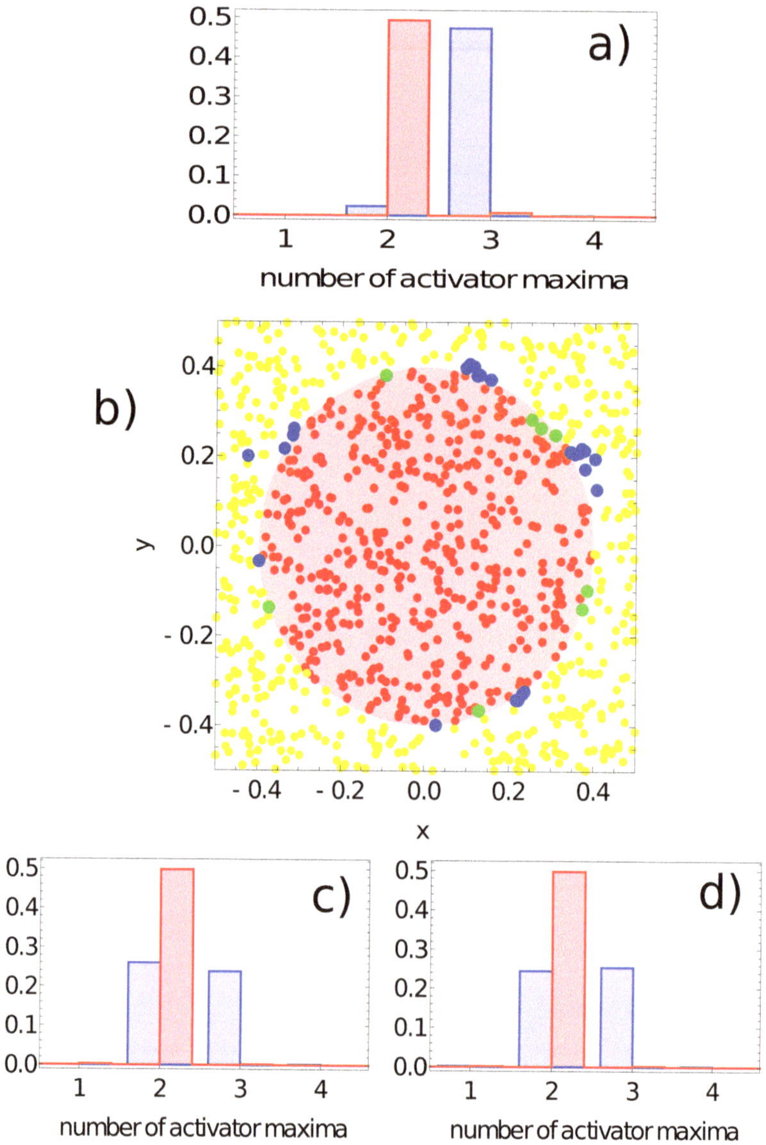

Figure 9. The answer of the network defined by the parameters listed in the first line of Table 1 to the records of the training dataset. Subfigures (**a**,**c**,**d**) are probability distributions of obtaining a given number of activator maxima on nodes #1, #2 and #3, respectively. The red bars correspond to points inside the red area; the blue bars refer to points outside the red area. Subfigure (**b**) illustrates correctly (yellow and red) and incorrectly (green and blue) classified points of the training dataset when node #1 is used as the output.

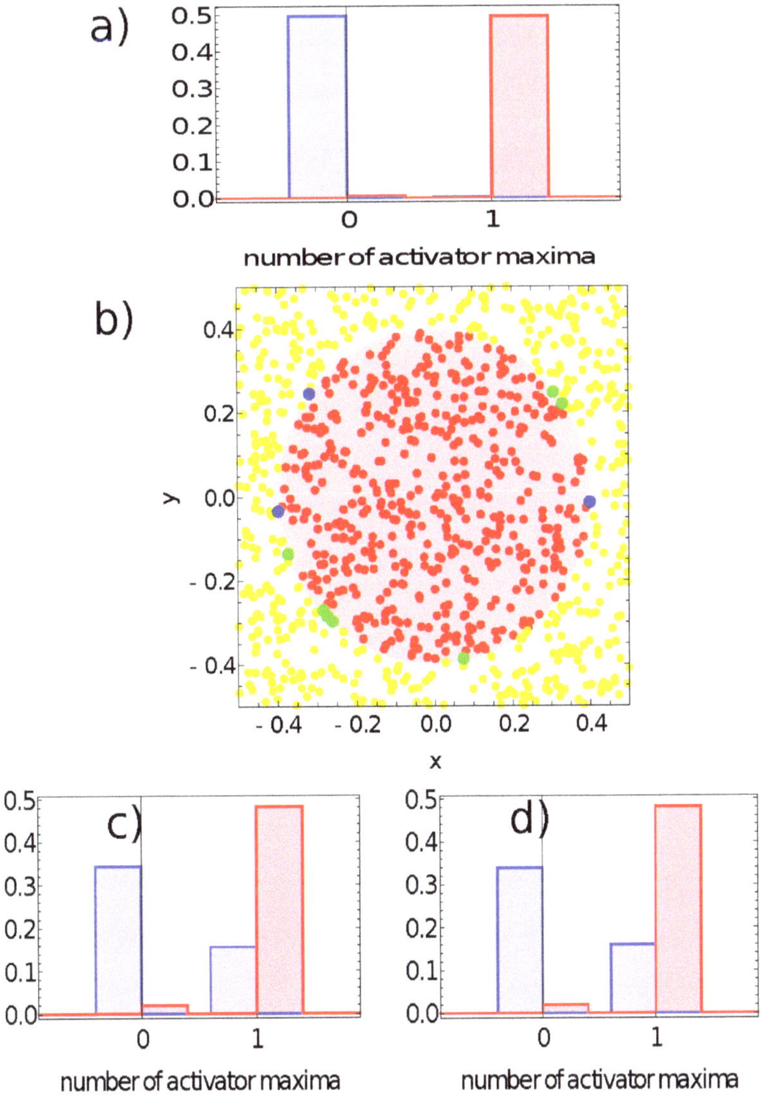

Figure 10. The answer of the network defined by the parameters listed in the second line of Table 1 to the records of the training dataset. Subfigures (**a**,**c**,**d**) are probability distributions of obtaining a given number of activator maxima on nodes #1, #2 and #3, respectively. The red bars correspond to points inside the red area; the blue bars refer to points outside the red area. Subfigure (**b**) illustrates correctly (yellow and red) and incorrectly (green and blue) classified points of the training dataset when node #1 is used as the output.

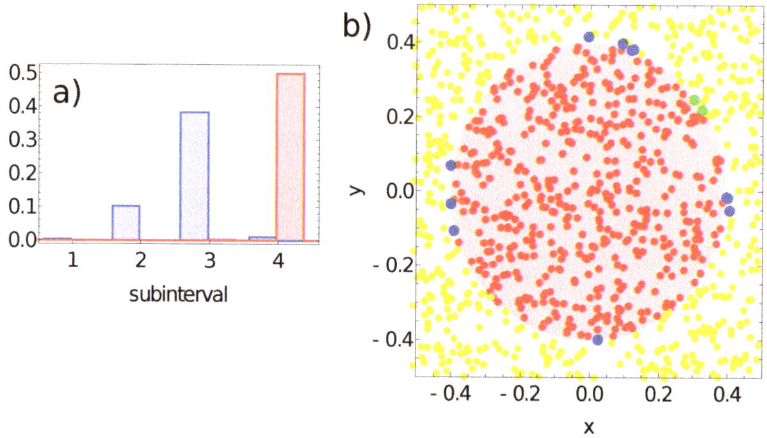

Figure 11. The answer of the network defined by the parameters listed in the third line of Table 1 to the records of training dataset. (**a**) The probability distribution of obtaining the value of $J_u = \int_0^{t_{max}} u_1(t)dt$ in the intervals $[k * 0.025, (k+1) * 0.025)$ for $k \in \{1, 2, 3, 4\}$. The red bars correspond to points inside the red area; the blue bars refer to points outside the red area. Subfigure (**b**) illustrates correctly (yellow and red) and incorrectly (green and blue) classified points of the training dataset.

Figure 9 shows the probability distribution of the number of activator maxima for the network defined by the parameters listed in the first line of Table 1. The red and blue bars correspond to points located in the sun and outside it, respectively. On nodes #2 and #3, we observe a large number of records representing points both in and out of the sun area that produced two maxima of the activator (cf. Figure 9c,d). On the other hand, for node #1, the number of activator maxima is two for most of the records representing points inside the sun area, and three maxima are observed for a large majority of points outside it. Therefore, node #1 is considered the output, and the classification rule is:

- If one or two maxima are observed, then the record represents a point in the sun area of the training dataset;
- If three or more maxima are observed, then the record represents a point outside the sun area.

Such a rule incorrectly classifies 24 points located outside the sun and 8 points located inside the sun area. Therefore, the accuracy of this network with this classification rule on the training dataset is 96.8%. The geometry of correctly and incorrectly classified records of the training database is illustrated in Figure 9b. In this figure, as well as on similar figures for the other considered classifiers, the red dots represent points inside the sun area that are correctly recognized by the network. Similarly, the yellow dots represent points outside the sun area that are correctly recognized by the network. The green dots are points located within the sun area, but the network classifies them as belonging to the white region. Finally, the blue dots are misclassified points located in the white region. In order to see the errors more clearly, the dots corresponding to incorrectly classified points are larger than those representing correct answers. The location of errors at the edge of the red area is non-homogeneous and suggests the existence of regions where errors dominate.

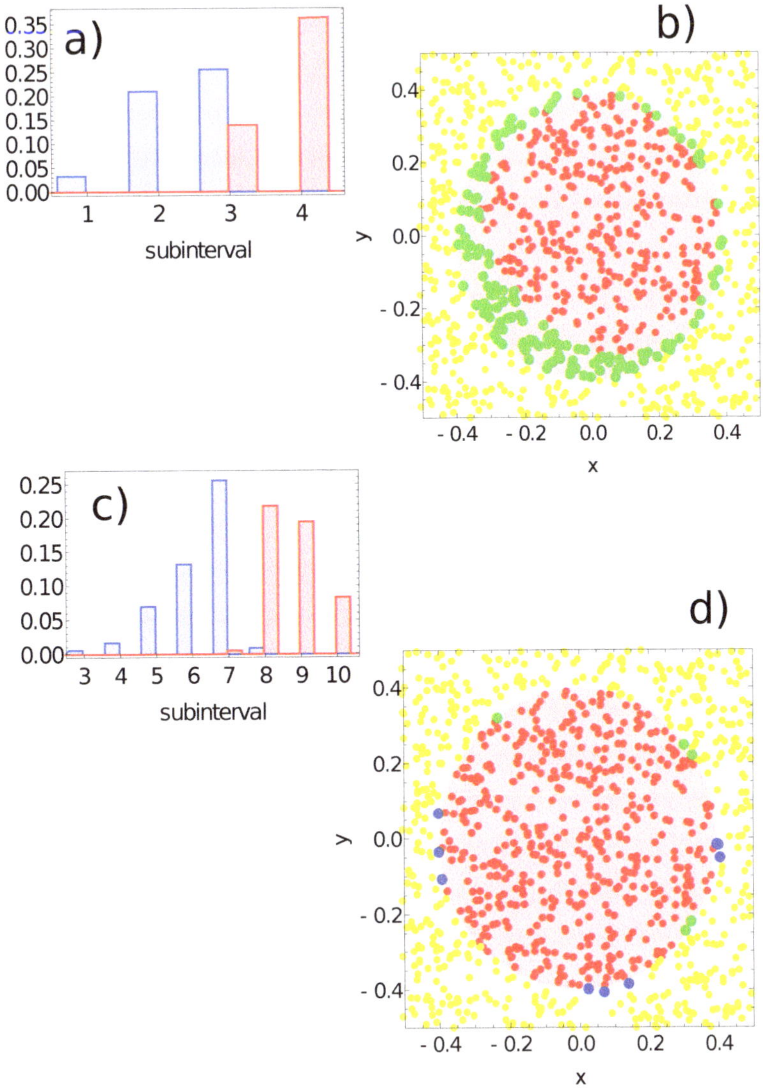

Figure 12. The answer of the network defined by the parameters listed in the fourth line of Table 1 to the records of training dataset. (**a**) The probability distribution of obtaining the value of $J_v = \int_0^{t_{max}} v_1(t)dt$ in the intervals $[k * 0.025, (k+1) * 0.025)$ for $k \to \{1,2,3,4\}$. The red bars correspond to points inside the red area; the blue bars refer to points outside the red area. Subfigure (**b**) illustrates correctly (yellow and red) and incorrectly (green) classified points of the training dataset. As in (**a**,**b**) but for node #3: (**c**) the probability distribution of obtaining the value of $J_u = \int_0^{t_{max}} v_3(t)dt$ in the intervals $[k * 0.025, (k+1) * 0.025)$ for $3 \leq k \leq 10$. Subfigure (**d**) illustrates correctly (yellow and red) and incorrectly (green and blue) classified points of the training dataset.

Similar results, but for the network defined by the parameters listed in the second line of Table 1, are shown in Figure 10. Again, node #1 is the network output because on nodes #2 and #3, we observe one activator maximum for a large number of records representing points located both in and out of the sun area (cf. Figure 10c,d). For node #1 (Figure 10a), most records representing points inside the sun area produced a single maximum, and for most points located outside the sun, no maximum was observed (cf. Figure 6b). Therefore, node #1 is the output, and the classification rule is:

- If a single maximum of the activator is observed, then the record represents a point in the sun area of the training dataset;
- If we record no maxima, then the processed data represent a point outside the sun area.

Such a rule incorrectly classifies just three points located outside the sun and seven points located inside the sun area. The accuracy of the second considered network with the classification rule listed above is 99% on the training dataset. The geometry of correctly and incorrectly classified records of the training database is illustrated in Figure 10b. All incorrectly classified points are located close to the boundary between the sun and the white region. Let us notice that the accuracy is 2% higher than for the previous network, and the only difference between these networks is the character of oscillations (cf. Figure 2b,c). This result suggests that "softer" oscillations may give networks with higher accuracy. This is the reason why the Oregonator Model II was used to investigate the alternative methods for relating the time evolution of concentration on a selected node with the network output.

Figure 11 presents results for the network defined by the parameters listed in the third line of Table 1. The network output is defined as: $J_u = \int_0^{t_{max}} u_j(t)dt$, where j denotes the output node. The range of J_u values is divided into subintervals I_k with length 0.025 as follows: $I_k = [k * 0.025, (k+1) * 0.025)$ for $k \in 1, 2, 3, 4$. Figure 11a illustrates the probability distribution of J_u values for output node #1. The classification rule can be formulated as follows:

- If the value of $J_u \geq 0.1$, then the record represents a point in the sun area of the training dataset;
- If the value of $J_u < 0.1$, then the record represents a point outside the sun area.

Such a rule incorrectly classifies just 2 points located inside the sun area and 11 points located outside it. All of the misclassified points are located close to the sun edge (cf. Figure 11b). The accuracy of the third of the considered networks with the above classification rule is 98.7% on the training database.

Keeping in mind the problem symmetry, we expect that node #1 should be the output. This was confirmed in the first three considered networks. However, node #1 gives poor accuracy for the network defined by the parameters listed in the fourth line of Table 1. For this network, the output is defined as: $J_v = \int_0^{t_{max}} v_j(t)dt$, where j denotes the output node. As for the previously discussed network, the range of J_v values is divided into subintervals I_k. Figure 12a illustrates the probability distribution of J_v values for output node #1. The records representing points in the sun area and those located outside it return values of J_v in the third subinterval I_3. Figure 12b shows the distribution of correctly and incorrectly classified records. For the majority rule derived from Figure 12a, all points located outside the sun are correctly classified, but there are many points inside the sun that are classified as located outside. Figure 12c illustrates the probability distribution of J_v values for output node #3. We observe a nice separation of records representing points in the sun area from those located outside. On the basis of this result, we formulate the following classification rule:

- If the value of $J_v \geq 0.2$, then the record represents a point in the sun area of the training dataset;
- If the value of $J_v < 0.2$, then the record represents a point outside the sun area.

Such a rule incorrectly classifies just five points located inside the sun area and nine points located outside it. All of the misclassified points are located close to the sun's edge (cf. Figure 12d). The accuracy of the fourth considered network with the above classification rule is 98.6% on the training database. Network symmetry is reflected by high classification accuracy based on the inhibitor's time evolution on node #2. If we use this node as the output, we get classification accuracy of 97.9%. It can be expected that the difference between the accuracy with node #2 versus node #3 as the output is related to the choice of the training dataset, and in a perfectly balanced choice, both of these nodes should have the same accuracy.

4. Discussion and Conclusions

In the previous section, I demonstrated that three-node networks can be optimized to determine the color of a point in the Japanese flag with high accuracy. However, the accuracy was achieved on the training dataset, raising the question of whether the results shown in Figures 9–12 are general, or if they reflect correlations inevitable in the small set of randomly selected points that formed the training database.

To produce a stronger argument for the accuracy of optimized networks, I verify them on a large test dataset that contained 100,000 records. The points are generated randomly, and the test dataset includes 49,916 records corresponding to points inside the sun area and 50,084 points outside it.

The answers of the networks defined by the parameters listed in Table 1 to the records of the testing dataset are presented in Figure 13. Red and yellow dots represent correctly classified points located inside and outside the red area of the flag. To reduce the figure size, the number of points is limited to 10,000. The green color marks points located in the red area that are wrongly classified as being located in the white part. The blue color denotes points from outside that are classified as belonging to the sun area. In all cases, the classification is accurate, and the flag is nicely represented.

Figure 13a shows the results for the networks defined by the parameters listed in the first line of Table 1. As suggested by the results in Figure 9, the number of activator maxima on node #1 is used as the output. The majority rule is:

- If one or two maxima are observed, then the point is within the sun area;
- If three or more maxima are observed, then the point is outside the sun area.

Such a rule incorrectly classifies 3216 points located outside the sun (blue in Figure 13a) and 1111 points located inside the sun area (green in Figure 13a). Therefore, the accuracy of this classifier is 95.668%. The sun area considered in this paper is more symmetrically oriented in the coordinate system than the Japanese flag located in the unit square $[0,1] \times [0,1]$ studied in [47]. For such a flag location the optimized classifier is characterized by: $t_{max} = 20.23$, $t_{start} = 3.78$, $t_{end} = 12.10$ and $t_{illum}(3) = 6.37$. Its accuracy tested on 100,000 records is 95.145%. However, to obtain this result, it was assumed that the values of parameter α_j were identical for all nodes ($= 0.849$). Moreover, the values of $\beta_{i,j} = 0.251$ were the same for all pairs of nodes. These assumptions definitely decreased the accuracy of the network reported in [47] if compared to the one defined in the first row of Table 1. Therefore, I expect that the accuracy of the three-oscillator classifier that codes the output in the number of activator maxima on a selected node does not systematically depend on the location of the flag in the coordinate system. The points classified as the sun (red and blue Figure 13a) form a characteristic horned disk already observed for a three-node classifier optimized to recognize the Japanese flag in the unit square $[0,1] \times [0,1]$ [47]. The fact that the horned shape is repeated by optimization using a training database containing points with coordinates in different ranges suggests that this shape is related to the parameters of the Oregonator Model I.

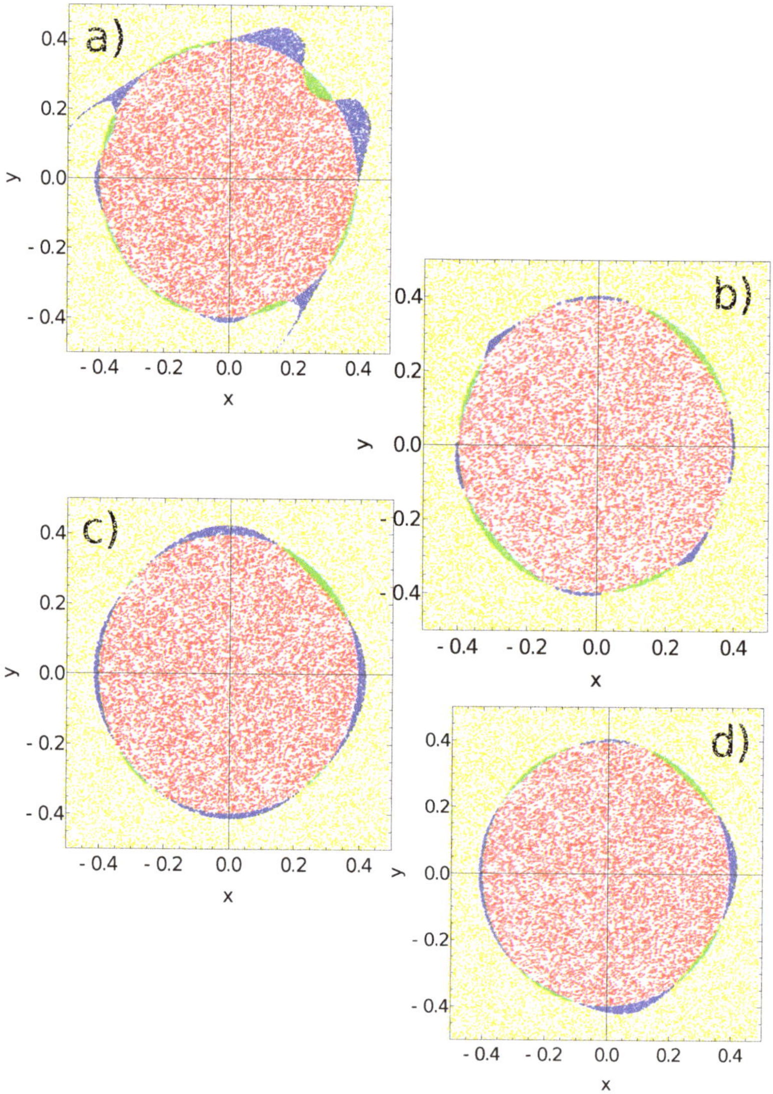

Figure 13. The answer of networks defined by the parameters listed in Table 1 to a testing dataset of 100,000 records. Yellow and red points are classified correctly. The network gives a wrong answer on points marked green (they belong to the sun but are classified as located outside it) and points marked blue (they are located outside the sun but are classified as belonging to the red area). Subfigures (**a–d**) correspond to networks with parameters listed in lines 1–4 of Table 1, respectively. The accuracy of these networks is (**a**) 0.957, (**b**) 0.984, (**c**) 0.976 and (**d**) 0.979, respectively.

The horned shape is not observed if one still uses the number of activator maxima as the output but considers another set of Oregonator parameters (Model II, (cf. Figure 13b)). A trace of undeveloped horns can be seen on the decreasing diagonal, but the sun disk is well represented. The classification rule (cf. Figure 10a) incorrectly predicts the color for 446 points located outside the sun (blue in Figure 13b) and 1192 points located inside the sun area (green in Figure 13b). The classifier accuracy is 98.362%. This result shows that the change in the oscillator model can improve the accuracy and modify the geometry of incorrectly classified points.

Figure 13c,d represent classifiers where the integrals of activator and inhibitor observed on the output node were used as the output. The analysis of output related to activator concentration is shown in Figure 13c. The points inside the sun were those for which $J_u \geq 0.1$. This rule incorrectly predicts the color for 2021 points located outside the sun (blue in Figure 13c) and only 401 points located inside the sun area (green in Figure 13c). The corresponding accuracy is 97.578%.

In the case of output related to inhibitor concentration (Figure 12c,d), the points inside the sun were those for which $J_v \geq 0.2$. This rule incorrectly predicts the color for 1409 points located outside the sun (blue in Figure 13d) and only 691 points located inside the sun area (green in Figure 13d). The corresponding accuracy is 97.900%. The use of node #3 as the output is suggested due to its having the highest accuracy obtained on the training dataset. However, slightly higher accuracy (97.963%) is obtained if node #2 acts as the output. This indicates that even for such a simple problem as the Japanese flag one, the size of the training dataset is important to obtain the correct rule of classification.

The comparison between the accuracy of described classifiers shows that the method based on the integration of activator or inhibitor concentration is as good as the output represented by the number of the corresponding maxima. In the method of maxima counting, there was a magic parameter—the threshold value for acceptance of maximum. The method based on integration does not need to involve such additional parameters to obtain the result and seems to be a promising candidate for future investigation on the computing potential of networks composed of interacting chemical oscillators.

Funding: This research was funded by the Institute of Physical Chemistry as the support for the research activity of the Laboratory of Complex Systems and Chemical Processing of Information.

Institutional Review Board Statement: Not applicable.

Informed Consent Statement: Not applicable.

Data Availability Statement: All presented data are available from the author.

Conflicts of Interest: The author declares no conflict of interest.

Abbreviation

The following abbreviation is used in this manuscript:

BZ Belousov–Zhabotinsky

References

1. Haken, H. *Brain Dynamics*; Springer Series in Synergetics; Springer: Berlin/Heidelberg, Germany, 2002.
2. Moore, G.E. Cramming More Components onto Integrated Circuits. *Electronics* **1965**, *38*, 114–117. [CrossRef]
3. Waldrop, M.M. The chips are down for Moore's law. *Nature* **2016**, *530*, 144–147. [CrossRef] [PubMed]
4. Feynman, R.P.; Hey, T.; Allen, R. *Feynman Lectures on Computation*; CRC Press: Boulder, CO, USA, 2000; ISBN 978-0738202969.
5. Adamatzky, A. (Ed.) *Advances in Uncnoventional Computing*; Springer: Cham, Switzerland, 2018; Volume 2, ISBN 978-3-319-33920-7.
6. Hjelmfelt, A.; Weinberger, E.D.; Ross, J. Chemical implementation of finite-state machines. *Proc. Natl. Acad. Sci. USA* **1992**, *89*, 383–387. [CrossRef] [PubMed]
7. Toth, A.; Showalter, K. Logic gates in excitable media. *J. Chem. Phys.* **1995**, *103*, 2058–2066. [CrossRef]
8. Steinbock, O.; Kettunen, P.; Showalter, K. Chemical wave logic gates. *J. Phys. Chem.* **1996**, *100*, 18970–18975. [CrossRef]
9. Adamatzky, A.; De Lacy Costello, B. Experimental logical gates in a reaction diffusion medium: The XOR gate and beyond. *Phys. Rev. E* **2002**, *66*, 046112. [CrossRef] [PubMed]
10. Gorecki, J.; Gorecka, J.N.; Igarashi, Y. Information processing with structured excitable medium. *Nat. Comput.* **2009**, *8*, 473–492. [CrossRef]
11. Magri, D.C.; Brown, G.J.; McClean, G.D.; de Silva, A.P. Communicating chemical congregation: A molecular AND logic gate with three chemical inputs as a "lab-on-a-molecule" prototype. *J. Am. Chem. Soc.* **2006**, *128*, 4950–4951. [CrossRef]
12. de Silva, A.P.; Uchiyama, S. Molecular logic and computing. *Nat. Nanotechnol.* **2007**, *2*, 399–410. [CrossRef]
13. Yoshikawa, K.; Motoike, I.N.; Ichino, T.; Yamaguchi, T.; Igarashi, Y.; Gorecki, J.; Gorecka, J.N. Basic information processing operations with pulses of excitation in a reaction-diffusion system. *Int. J. Unconv. Comput.* **2009**, *5*, 3–37.

14. McKinney, B.O.F.; Daly, B.; Yao, C.; Schroeder, M.; de Silva, A.P. Consolidating Molecular Logic with New Solid-Bound YES and PASS 1 Gates and Their Combinations. *ChemPhysChem* **2017**, *18*, 1760. [CrossRef] [PubMed]
15. Adamatzky, A.; De Lacy Costello, B.; Asai, T. *Reaction-Diffusion Computers*; Elsevier: New York, NY, USA, 2005.
16. Lin, X.; Liu, Y.; Deng, J.; Lyu, Y.; Qian, P.; Lia, Y.; Wang, S. Multiple advanced logic gates made of DNA-Ag nanocluster and the application for intelligent detection of pathogenic bacterial genes. *Chem. Sci.* **2018**, *9*, 1774–1781. [CrossRef]
17. Belousov, B.P. *Collection of Short Papers on Radiation Medicine*; Medgiz: Moscow, Russia, 1959; pp. 145–152.
18. Zhabotinsky, A.M. Periodic liquid phase reactions. *Proc. Acad. Sci. USSR* **1964**, *157*, 392–395.
19. Field, R.J.; Burger, M. *Oscillations and Traveling Waves in Chemical Systems*; Field, R.J., Burger, M., Eds.; Wiley: New York, NY, USA, 1985; ISBN 0-471-89384-6.
20. Epstein, I.R.; Pojman, J.A. *An Introduction to Nonlinear Chemical Dynamics: Oscillations, Waves, Patterns, and Chaos*; Oxford University Press: New York, NY, USA, 1998; ISBN 9780195096705.
21. Motoike, I.; Yoshikawa, K. Information Operations with an Excitable Field. *Phys. Rev. E* **1999**, *59*, 5354–5360. [CrossRef] [PubMed]
22. Gorecka, J.N.; Gorecki, J. Multiargument logical operations performed with excitable chemical medium. *J. Chem. Phys.* **2006**, *124*, 084101. [CrossRef]
23. Gentili, P.L.; Giubila, M.S.; Germani, R.; Romani, A.; Nicoziani, A.; Spalletti, A.; Heron, B.M. Optical Communication among Oscillatory Reactions and Photo-Excitable Systems: UV and Visible Radiation Can Synchronize Artificial Neuron Models. *Angew. Chem. Int. Ed.* **2017**, *56*, 7535. [CrossRef] [PubMed]
24. Kuhnert, L. A new optical photochemical memory device in a light-sensitive chemical active medium. *Nature* **1986**, *319*, 393–394. [CrossRef]
25. Kádár, S.; Amemiya, T.; Showalter, K. Reaction Mechanism for Light Sensitivity of the $Ru(bpy)_3^{2+}$-Catalyzed Belousov–Zhabotinsky Reaction. *J. Phys. Chem. A* **1997**, *101*, 8200–8206. [CrossRef]
26. Agladze, K.; Aliev, R.; Yamaguchi, T. Chemical Diode. *J. Phys. Chem.* **1996**, *100*, 13895–13897. [CrossRef]
27. Motoike, I.N.; Yoshikawa, K.; Iguchi, Y.; Nakata, S. Real-Time Memory on an Excitable Field. *Phys. Rev. E* **2001**, *63*, 036220. [CrossRef] [PubMed]
28. Gorecki, J.; Gorecka, J.N. On mathematical description of information processing in chemical systems. In *Mathematical Approach to Nonlinear Phenomena: Modeling, Analysis and Simulations*; GAKUTO International Series: Mathematical Sciences and Applications; Gakkotosho: Tokyo, Japan, 2005; Volume 23, pp. 73–90. ISBN 4762504327.
29. Adleman, L.M. Molecular Computation of Solutions to Combinatorial Problems. *Science* **1994**, *266*, 1021–1024. [CrossRef] [PubMed]
30. Calude, C.S.; Păun, G. *Computing with Cells and Atoms*; Taylor & Francis Publisher: London, UK, 2002.
31. Steinbock, O.; Toth, A.; Showalter, K. Navigating complex labyrinths—Optimal paths from chemical waves. *Science* **1995**, *267*, 868–871. [CrossRef] [PubMed]
32. Agladze, K.; Magome, N.; Aliev, R.; Yamaguchi, T.; Yoshikawa, K. Finding the optimal path with the aid of chemical wave. *Physica D* **1997**, *106*, 247–254. [CrossRef]
33. Kuhnert, L.; Agladze, K.I.; Krinsky, V.I. Image processing using light-sensitive chemical waves. *Nature* **1989**, *337*, 244–247. [CrossRef]
34. Rambidi, N.G.; Maximychev, A.V. Towards a biomolecular computer. Information processing capabilities of biomolecular nonlinear dynamic media. *Biosystems* **1997**, *41*, 195–211. [CrossRef]
35. Szymanski, J.; Gorecka, J.N.; Igarashi, Y.; Gizynski, K.; Gorecki, J.; Zauner, K.P.; de Planque, M. Droplets with information processing ability. *Int. J. Unconv. Comput.* **2011**, *7*, 185–200.
36. Gorecki, J.; Gizynski, K.; Guzowski, J.; Gorecka, J.N.; Garstecki, P.; Gruenert, G.; Dittrich, P. Chemical Computing with Reaction-Diffusion processes. *Phil. Trans. R. Soc. A.* **2015**, *373*, 20140219. [CrossRef]
37. Guzowski, J.; Gizynski, K.; Gorecki, J.; Garstecki, P. Microfluidic platform for reproducible self-assembly of chemically communicating droplet networks with predesigned number and type of the communicating compartments. *Lab Chip* **2016**, *16*, 764–772. [CrossRef] [PubMed]
38. Buisman, H.J.; ten Eikelder, H.M.M.; Hilbers, P.A.J.; Liekens, A.M.L. Computing Algebraic Functions with Biochemical Reaction Networks. *Artif. Life* **2009**, *15*, 5–19. [CrossRef] [PubMed]
39. Hjelmfelt, A.; Weinberger, E.D.; Ross, J. Chemical implementation of neural networks and Turing machines *Proc. Nati. Acad. Sci. USA* **1991**, *88*, 10983–10987. [CrossRef] [PubMed]
40. Hjelmfelt, A.; Ross, J. Chemical implementation and thermodynamics of collective neural networks. *Proc. Nati. Acad. Sci. USA* **1992**, *89*, 388–391. [CrossRef] [PubMed]
41. Hjelmfelt, A.; Ross, J. Pattern recognition, chaos, and multiplicity in neural networks of excitable systems. *Proc. Natl. Acad. Sci. USA* **1994**, *91*, 63–67. [CrossRef] [PubMed]
42. Adamatzky, A.; Holley, J.; Bull, L.; De Lacy Costello, B. On computing in fine-grained compartmentalised Belousov–Zhabotinsky medium. *Chaos Solitons Fractals* **2011**, *44*, 779–790. [CrossRef]
43. Holley, J.; Adamatzky, A.; Bull, L.; De Lacy Costello, B.; Jahan, I. Computational modalities of Belousov–Zhabotinsky encapsulated vesicles. *Nano Commun. Netw.* **2011**, *2*, 50–61. [CrossRef]
44. Adamatzky, A.; Holley, J.; Dittrich, P.; Gorecki, J.; De Lacy Costello, B.; Zauner, K.-P.; Bull, L. On architectures of circuits implemented in simulated Belousov–Zhabotinsky droplets. *Biosystems* **2012**, *109*, 72–77. [CrossRef]

45. Gizynski, K.; Gorecki, J. Cancer classification with a network of chemical oscillators. *Phys. Chem. Chem. Phys.* **2017**, *19*, 28808–28819. [CrossRef]
46. Gizynski, K.; Gruenert, G.; Dittrich, P.; Gorecki, J. Evolutionary design of classifiers made of oscillators containing a nonlinear chemical medium. *MIT Evol. Comput.* **2017**, *25*, 643–671. [CrossRef]
47. Gorecki, J.; Bose, A. How Does a Simple Network of Chemical Oscillators See the Japaneseese Flag? *Front. Chem.* **2020**, *8*, 580703. [CrossRef]
48. Duenas-Diez, M.; Perez-Mercader, J. How Chemistry Computes: Language Recognition by Non-Biochemical Chemical Automata. From Finite Automata to Turing Machines. *iScience* **2019**, *19*, 514–526. [CrossRef]
49. Bose, A.; Gorecki, J. Computing With Networks of Chemical Oscillators and its Application for Schizophrenia Diagnosis. *Front. Chem.* **2022**, *10*, 848685. [CrossRef]
50. Field, R.J.; Noyes, R.M. Oscillations in chemical systems. IV. Limit cycle behavior in a model of a real chemical reaction. *J. Chem. Phys.* **1974**, *60*, 1877–1884. [CrossRef]
51. Sutthiopad, M.; Luengviriya, J.; Porjai, P.; Tomapatanaget, B.; Müller, S.C.; Luengviriya, C. Unpinning of spiral waves by electrical forcing in excitable chemical media. *Phys. Rev. E* **2014**, *89*, 052902. [CrossRef] [PubMed]
52. Tanaka, M.; Nagahara, H.; Kitahata, H.; Krinsky, V.; Agladze, K.; Yoshikawa, K. Survival versus collapse: Abrupt drop of excitability kills the traveling pulse, while gradual change results in adaptation. *Phys. Rev. E* **2007**, *76*, 016205. [CrossRef] [PubMed]
53. Gorecki, J.; Gorecka, J.N.; Adamatzky, A. Information coding with frequency of oscillations in Belousov–Zhabotinsky encapsulated disks. *Phys. Rev. E* **2014**, *89*, 042910. [CrossRef] [PubMed]
54. Gizynski, K.; Gorecki, J. Chemical memory with states coded in light controlled oscillations of interacting Belousov–Zhabotinsky droplets. *Phys. Chem. Chem. Phys.* **2017**, *19*, 6519–6531. [CrossRef]
55. Press, W.H.; Teukolsky, S.A.; Vetterling, W.T.; Flannery, B.P. *Numerical Recipes: The Art of Scientific Computing*; Cambridge University Press: New York, NY, USA, 2007; ISBN 978-0-521-88068-8.
56. MacKay, D.J.C. *Information Theory, Inference and Learning Algorithms*; Cambridge University Press: Cambridge, UK, 2003; ISBN 978-0-521-64298-9.
57. Cover, T.M.; Thomas, J.A. *Elements of Information Theory*; Wiley InterScience: New York, NY, USA, 2006; ISBN 978-0-471-24195-9.
58. Shannon, C.E. A Mathematical Theory of Communication. *Bell Syst. Tech. J.* **1948**, *27*, 379–423. [CrossRef]
59. Gorecki, J. Applications of Information Theory Methods for Evolutionary Optimization of Chemical Computers. *Entropy* **2020**, *22*, 313. [CrossRef]
60. Goldberg, D.E. *Genetic Algorithms in Search, Optimization and Machine Learning*; Addison-Wesley Longman Publishing Co., Inc.: Boston, MA, USA, 1989.

entropy

MDPI

Article

Effect of Ion and Binding Site on the Conformation of Chosen Glycosaminoglycans at the Albumin Surface

Piotr Sionkowski [1], Piotr Bełdowski [2,*], Natalia Kruszewska [2], Piotr Weber [3], Beata Marciniak [4] and Krzysztof Domino [1]

1 Institute of Theoretical and Applied Informatics, Polish Academy of Sciences, ul. Bałtycka 5, 44-100 Gliwice, Poland; piotr.sionkowski@gmail.com (P.S.); kdomino@iitis.pl (K.D.)
2 Institute of Mathematics and Physics, Bydgoszcz University of Science and Technology, 85-796 Bydgoszcz, Poland; nkruszewska@pbs.edu.pl
3 Institute of Physics and Applied Computer Science, Faculty of Applied Physics and Mathematics, Gdańsk University of Technology, ul. G. Narutowicza 11/12, 80-233 Gdańsk, Poland; piotr.weber@pg.edu.pl
4 Faculty of Telecommunications, Computer Science and Electrical Engineering, Bydgoszcz University of Science and Technology, 85-796 Bydgoszcz, Poland; beata.marciniak@pbs.edu.pl
* Correspondence: piotr.beldowski@pbs.edu.pl

Abstract: Albumin is one of the major components of synovial fluid. Due to its negative surface charge, it plays an essential role in many physiological processes, including the ability to form molecular complexes. In addition, glycosaminoglycans such as hyaluronic acid and chondroitin sulfate are crucial components of synovial fluid involved in the boundary lubrication regime. This study presents the influence of Na^+, Mg^{2+} and Ca^{2+} ions on human serum albumin–hyaluronan/chondroitin-6-sulfate interactions examined using molecular docking followed by molecular dynamics simulations. We analyze chosen glycosaminoglycans binding by employing a conformational entropy approach. In addition, several protein–polymer complexes have been studied to check how the binding site and presence of ions influence affinity. The presence of divalent cations contributes to the decrease of conformational entropy near carboxyl and sulfate groups. This observation can indicate the higher affinity between glycosaminoglycans and albumin. Moreover, domains IIIA and IIIB of albumin have the highest affinity as those are two domains that show a positive net charge that allows for binding with negatively charged glycosaminoglycans. Finally, in discussion, we suggest some research path to find particular features that would carry information about the dynamics of the particular type of polymers or ions.

Keywords: human serum albumin; hyaluronan; conformational entropy; dihedral angles; frequency distribution

Citation: Sionkowski, P.; Bełdowski, P.; Kruszewska, N.; Weber, P.; Marciniak, B.; Domino, K. Effect of Ion and Binding Site on the Conformation of Chosen Glycosaminoglycans at the Albumin Surface. *Entropy* **2022**, *24*, 811. https://doi.org/10.3390/e24060811

Academic Editor: Antonio M. Scarfone

Received: 11 May 2022
Accepted: 8 June 2022
Published: 10 June 2022

1. Introduction

Lubrication in natural joints is a complex multiscale process that involves interactions between constituents of articular cartilage and synovial fluid [1–3]. Although two main mechanisms—hydration repulsion and molecular synergies—have been found, the atomistic details on phenomena are still studied due to the lack of knowledge of how the cooperation between them results in facilitated lubrication [4–8]. Synovial fluid consists of up to 80% of water, macromolecular and small-molecular components.

The most recognizable components are: albumin, hyaluronan, phospholipids, γ-globulin and lubricin [9]. Human serum albumin (HSA) deserves special attention due to its binding and transporting properties of various compounds (fatty acids, ions: K^+, Na^+, Mg^{2+}, Ca^{2+} and many other molecules [10,11]). It was demonstrated in [12,13] that albumin and γ-globulin play an important role in lubrication. However, their influence on lubricating properties starts to be vital when taking into account their cooperation with other synovial fluid components. Locally positively charged sites of albumin favor interactions with the ionized carboxylic and sulfate groups in glycosaminoglycans (GAGs). Even though both macromolecules have a global negative charge under physiological conditions, it was shown that HSA could bind to negatively charged surfaces [14].

This study presents the analysis of the interactions between HSA and hyaluronic acid (HA)/ chondroitin-6-sulfate (CS6) in terms of the influence on conformations. As the topic is extensive, we are skipping here the conformation changes within albumin, focusing only on the GAGs. Such analysis of interaction between HSA and HA has been performed recently in [15]. HSA consists of a single chain of 585 amino acids, incorporating three homologous domains (I, II, and III) [16]. Domain I consists of residues 5–197, domain II includes residues 198–382, and domain III is formed from residues 383–569. Each domain comprises two subdomains termed A and B (IA; residues 5–107, IB; residues 108–197, IIA; residues 198–296, IIB; residues 297–382, IIIA; residues 383–494, IIIB; residues 495–569), these are depicted in Figure 1. Some of the domains are more prone to binding to GAGs than others; however, the binding map can alter under some conditions of various diseases [17,18]. The binding mechanism is mainly due to ionic, hydrogen bonding and hydrophobic interactions [17].

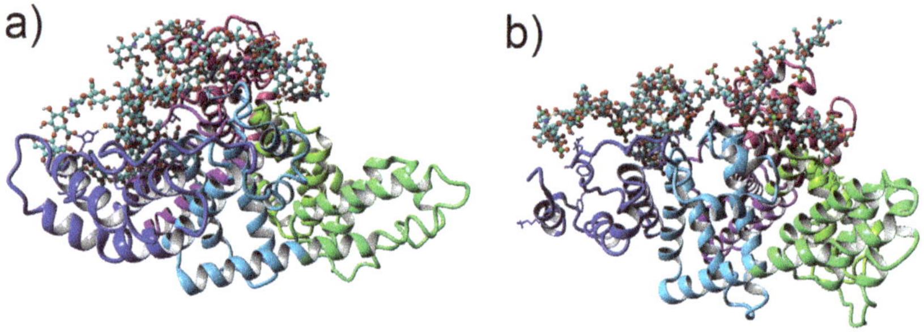

Figure 1. Structure of (**a**) HSA-HA, (**b**) HSA-CS6 complexes for highest affinity result in $CaCl_2$ solution (solution is transparent on the picture). HSA is depicted as ribbon (bottom parts of picture), and its domains are colored as follows: IA-pink, IB-violet, IIA-light green, IIB-green, IIIA-light blue, and IIIB-blue. HA and CS6 are depicted as ball-stick (top parts of picture). Light blue atoms represent carbon, dark blue nitrogen, red oxygen, green sulfur and white hydrogen. Snapshots was taken using YASARA software after 100 ns MD simulations [19].

GAGs are large complex carbohydrates. Depending on the monosaccharide types and the glycosidic bonds, GAGs can be divided into four groups: (i) hyaluronic acid, (ii) chondroitin sulfate, and dermatan sulfate (iii) heparan sulfate and heparin, and (iv) keratan sulfate. First, let us underline that HA is only a non-sulfated GAG. It is vital because sulfate groups in the GAG is one of the most crucial factors influencing the interaction map between a protein and the GAG [20]. Thus, two different examples of GAGs have been picked, which are essential components of synovial fluid, to study their conformational entropy while they are in the vicinity of the albumin: CS6 and HA. CS6 was shown to be an excellent material for bone regeneration as it is the main constituent of glycosaminoglycan in cartilage. CS6 is involved in bone homeostasis and in the coordination of osteoblastic cell attachment. Kim et al. investigated the role of chondroitin sulfate's negative charge on the binding of cations (e.g., calcium and phosphate) and showed that the hydroxyapatite crystal formation was enhanced, accelerating osteogenesis.

The analysis of the interaction between HSA and HA or CS6 in the presence of various species such as water and ions is a meaningful task; as such, interaction is closely related to synovial systems' unique properties [4]. The use of molecular modeling allows for researchers to evaluate the influence of various factors, such as the presence of ions and solvation on the properties of proteins, including their ability to bind ligands [21]. Following this, the present study aims to evaluate the effect of Na^+, Mg^{2+}, Ca^{2+} cations on the affinity of this two specific GAGs to HSA using: firstly docking, and secondly molecular dynamics

(MD) methods. The studies are focused on the description of changes in conformation of the GAGs in the vicinity of the HSA.

GAGs are important complexes that participate in many biological processes through the regulation of specific proteins. Hence, their secondary structure and stability are very important to study. Both of above-mentioned properties can be well quantified by conformational entropy. As conformational entropy, we understand the Shannon entropy computed for bivariate histograms of chosen pairs of dihedral angles (see, Section 2.1) [22]. Variations of such entropy are considered to measure important properties of the biochemical processes [23,24]. In this research, we also refer to the informative interpretation of entropy. Basically, one can compute this entropy by studying the motion of the molecule. There are various methods dedicated to characterize these motions (e.g., NMR relaxation methods [25], AFM-unfolding [26], and neutron spectroscopy [27]). Theoretical studies based on classical mechanics approach are a reasonable alternative but computationally demanding [28,29]. The intermediate approach, we follow, is the computation of conformational entropy from all-atom MD simulations, see [23,30,31].

In more detail, conformation description of the GAGs relays on the analysis of their structures bound to HSA domains in aqueous ionic solution. This analysis is carried out to check whether there are any differences in the conformation of the glycosidic linkages between each oligosaccharide monomer of the GAG, when the kind of the ion is changed in aqueous solution. The linkages are investigated basing on specific dihedral angles. In the present paper, conformational entropy is computed from the frequency distribution of those angles' values. We anticipate that those angles determine important characteristics, such as shape and stiffness. As the conformational entropy is calculated from the distribution of the angles, it is expected to be a crucial feature (enclosing the quantitative description in one numerical value).

2. Materials and Methods

We have performed all-atom simulations of the two model biosystems (one is HSA with HA and the other is HSA with CS6) in aqueous ionic solution. First, a molecular docking procedure has been executed to obtain preliminary information on the stability of the structure and to find the most energetically optimal places where each GAG attaches to the HSA. Next, energetically best-docked structures (sorted from the strongest connection to the weakest connected), with added water solution of chosen ions (Na^+, Mg^{2+}, Ca^{2+} and Cl^-), have been subjected to MD simulations.

Chemical structures of HA and CS6 were obtained from Pubchem and modified to obtain chains of around 8 kDa (24 units). This modification relied on connecting units of selected GAGs until polymers of desired length were obtained. To acquire the most stable complexes, we docked GAG ligand (HA or CS6) to HSA using the VINA method [32] with their default parameters and point-charge force field [33] initially assigned according to the AMBER14 force field [34] (the HA molecule was parametrized by applying the GLYCAM06 force field [35]). Then, we damped the system to mimic the less polar Gasteiger charges used to optimize the AutoDock scoring function. The simulation was done with the YASARA molecular modeling program [19]. In each case (HA and CS6), 10 of the best distinctive complexes which differs in the position of GAG docked to HSA (best complexes means the complexes of the highest energy of binding and RMSD of complexes from the best binded complex values, which were computed by VINA) with −10 kcal/mol free energy of binding prepared for MD simulation.

MD simulations of HSA (PDB code: 1e78) with GAG have been run with YASARA software. Optimization of the hydrogen bonding network was included in the setup to increase the solute stability and a pK_a prediction (based on a Henderson–Hasselbalch equation) to fine-tune the protonation states of the protein residues at the given pH = 7.4 [36,37]. Optimization was done in three main steps: (a) pKa prediction was included to consider the influence of the pH on the hydrogen bonding network, (b) nonstandard amino acids and ligands were fully accounted for with the help of a chemical knowledge library in SMILES format, and (c) the use of the SCWRL algorithm allows for finding the globally optimal solution. In the case of the HSA-HA system, the complex has been immersed in one of the three aqueous 0.9% salt solutions NaCl, $CaCl_2$ or $MgCl_2$. In the case of HSA-CS6, it was 2% water solution of the same salts in pH = 7.0. After necessary minimization of the model system to remove clashes, the simulation was run for 100 ns using the AMBER14 force field [34] for the HSA, GLYCAM06 [35] for HA and CS6, and TIP3P for water. The van der Waals forces' cut-off distance was set to 10 Å [38]. The particle Mesh Ewald algorithm was used for computing long-range interactions (e.g., electrostatic interactions) [39]. Simulations were performed under following conditions: temperature 310 K and pressure of 1 atm (NPT ensemble) [37]. Periodic boundary conditions were applied to a box of size roughly equal to 120 $Å^3$. Berendsen barostat and thermostat were used to maintain constant pressure and temperature (relaxation time of 1 fs) [40]. The equations of motions were integrated with multiple time steps of 1.25 fs for bonded interactions and 2.5 fs for non-bonded interactions. In the considered simulations, the time step between saved states of the system equals 100 ps. Thus, the time series for 100 ns of simulations obtained 1000 save points. Snapshots of the two complexes after 100 ns of MD simulation have been presented in Figure 1.

All analyses and computation have been performed using YASARA and in-house written data processing programs in Python 3.8 [41].

2.1. Backbone Angles Determination

The method of entropy calculation, used in this study, relies on computation of the frequency distribution of the backbone's dihedral angles (Φ,Ψ), usually presented in the form of a Ramachandran-type plot [31,42]. It provides a simplistic view of the conformation of a molecule by clustering angles (Φ,Ψ) into district regions (bear in mind that there are two sets of such pairs of angles, which will be discussed later).

The CS6 consists of glucuronic acid (GlcA) and galactosamine (sulfated at C-6 atom of galactosamine; GalNAc), while HA consists of glucuronic acid (GlcA) and acetylglucosamine (GlcNAc). Linkages between the two monosaccharides are as follows: in the case of CS6, it is [4)-β-GlcA-(1⟶3)-β-GalNAc-)1⟶], and for HA: [4)-β-GlcA-(1⟶3)-β-GlcNAc-)1⟶] (see, Figure 2). Using abbreviation G for GlcA and N for GlcNAc (in HA) or GalNAc (in CS6), we can depict the glycosaminoglycans as linear heteropolysaccharide chains consisted of repeating disaccharide units [31]. In the presented study, the chains consist of NG units repeated 24 times. Two sets of torsion angles describe the conformations around the glycosidic linkages: Φ_{1-4} and Ψ_{1-4} (N-G linkage), and Φ_{1-3} and Ψ_{1-3} (G-N linkage) [31]. Thus, in the GAG chains, there are 24 of Φ_{1-4} and Ψ_{1-4} angles, and 23 Φ_{1-3} and Ψ_{1-3}. These angles can be written as follows:

$$\begin{aligned} \Phi_{1-4} &= \text{O5(N)-C1(N)-O1(N)-C4(G)}, \\ \Psi_{1-4} &= \text{C1(N)-O1(N)-C4(G)-C5(G)}, \\ \Phi_{1-3} &= \text{O5(G)-C1(G)-O1(G)-C3(N)}, \\ \Psi_{1-3} &= \text{C1(G)-O1(G)-C3(N)-C4(N)}. \end{aligned} \tag{1}$$

All dihedral angles in Equation (1) have been presented in Figure 2. The available conformational space of the GAGs' chains depends mainly on the two torsion angles. As mentioned before, a goal of the present study is to investigate how various features of the system (ions, HA vs. CS6) affect the frequency distribution of Φ and Ψ torsion angles. As the informative feature of above-mentioned frequency distribution, conformational entropy has been used [24,31]. Note that, in information theory, entropy is the measure of the information tied directly to the variable (univariate or multivariate) [22,43]. Using this analogy, we can analyse which pairs of dihedral angles are more or less informative for specific GAGs and ions.

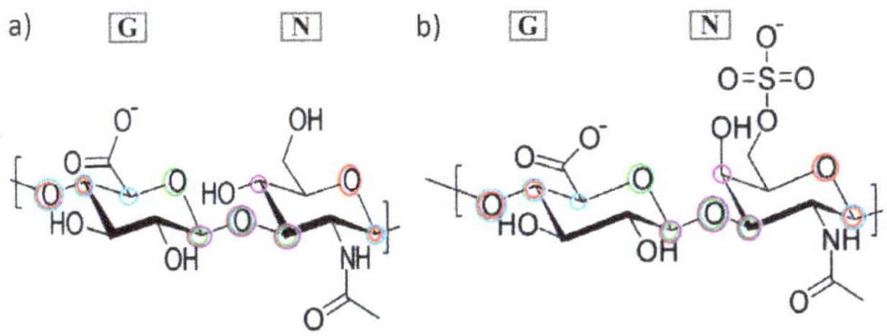

Figure 2. Structures of (**a**) HA and (**b**) CS6 with dihedral angles: Φ_{1-4}—red circles, Ψ_{1-4}—blue circles (N-G linkage), Φ_{1-3}—green circles, Ψ_{1-3}—violet circles (G-N linkage).

2.2. Entropy Calculation

For each pair of subsequent mers of GAG's chain, the time series (containing 1000 points) of the dihedral angles have been obtained. As described in Section 2.1, following pairs of these angles: Φ_{1-4} vs. Ψ_{1-4} and Φ_{1-3} vs. Ψ_{1-3} have been analyzed [31]. 2D histograms have been calculated from angles values of all subsequent pairs of mers (and all time steps). Histograms of the stronger bound structures (from the 10 picked as described in Section 2), for different ionic solutions, have been presented in Figure 3 for CS6 and in Figure 4 for HA.

Following an approach described in [44] (see Figure 8 therein), the conformation entropy has been computed from the 2D histograms of pairs of angles. In more detail, from each histogram, the Shannon entropy [22,23,30] has been computed using the formula:

$$S = -R_0 \sum_{i,j} p_{i,j} \log(p_{i,j}). \tag{2}$$

Here, $p_{i,j}$ is the empirical probability of the first angle being in the ith bin and the second angle being in the jth bin, and $R_0 = 8.314\ \mathrm{J{\cdot}K^{-1}{\cdot}mol^{-1}}$ is the (scaling) gas constant.

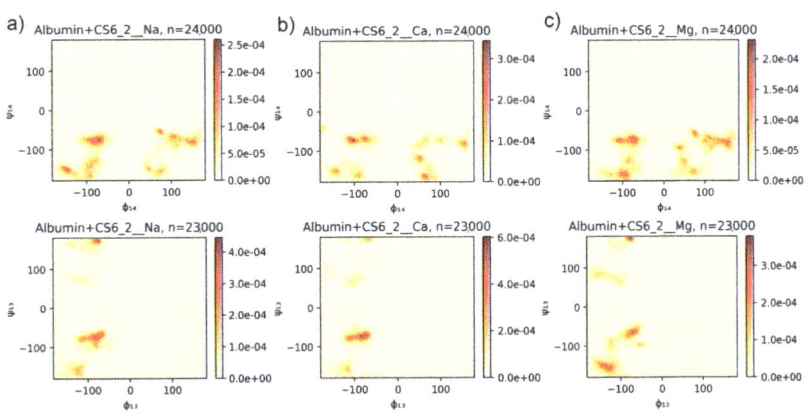

Figure 3. Normalized histograms of values of angles (Φ_{1-4}, Ψ_{1-4}) (top) and (Φ_{1-3},Ψ_{1-3}) (bottom) for main chain CS6 with (**a**) Na^{+}, (**b**) Ca^{2+}, and (**c**) Mg^{2+}. Angles were taken from the whole time series of the YASARA simulation. The symbol n is a number of angles' pairs and is equal to number of angles type (24 for angles 1, 4 and 23 for angles 1, 3, cf. Section 2.1) multiplied by number of time points (1, 000).

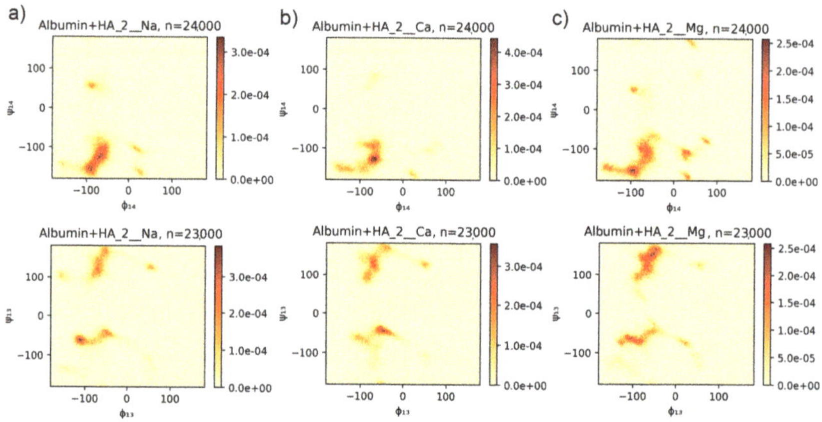

Figure 4. Normalized histograms of values of angles (Φ_{1-4}, Ψ_{1-4}) (top) and (Φ_{1-3},Ψ_{1-3}) (bottom) for main chain HA with (**a**) Na^+, (**b**) Ca^{2+}, and (**c**) Mg^{2+}. Angles were taken from the whole time series of the YASARA simulation. The symbol n is a number of angles' pairs and is equal to number of angles type (24 for angles 1, 4 and 23 for angles 1, 3, cf. Section 2.1) multiplied by number of time points (1, 000).

3. Results

Some HSA segments are more prone to creating intermolecular interactions than others. The complexity of protein–GAG interactions is in part caused by the conformational flexibility of the GAG's chain. An affinity of the HSA to GAGs, firstly tested by docking method, has been present in Table 1 sorted by binding energy. While the docking method relies on adjusting a ligand to a receptor in crystal form, then putting the complex into a water solution changes the intermolecular interactions map. After equilibration and 100 ns of MD simulations, the binding energy changed, and the order of best-bound complexes changed. In the case of CS6, the new order depended on added ions, and its value (averaged over three realizations' binding energies with a different salt added) has been written in the first column of Table 1 in the brackets. HSA binding sites did not change much during the MD simulations. However, the number of interactions such as hydrophobic-polar, hydrogen bonds, ionic, and bridges have changed.

Table 1. Binding ranks of HSA-CS6 complexes. The first column contains: rank after docking (averaged rank after MD simulations). The strongest connected domains are marked in bold letters.

HSA-CS6 Complex Rank	HSA Binding Sites
1(5)	IA-IIA-IIIA-**IIIB**
2(1)	IA-IB-IIA-**IIIA**-IIIB
3(8)	IA-**IIA**
4(6)	IA-IIA-IIIA-**IIIB**
5(7)	IA-IIA-IIIA-**IIIB**
6(2)	IB-IIIA-**IIIB**
7(3)	**IA**-IIA-IIB-IIIA
8(10)	**IA**-IB
9(4)	IB-IIIA-**IIIB**
10(9)	IIA-**IIB**

A closer analysis of the interactions for HSA-CS6 complexes will be the subject of another study, similar to the ones performed for HSA-HA complexes [15]. In the present paper, we are focused only on the conformational entropy of the GAGs chains. In the case of CS6, the most stable turned out realization was number 2, thus the situation when CS6 wrapped around the HSA and bound to IA-IB-IIA-IIIA-IIIB domains had the strongest binding to IIIA. No matter what ions have been added to the system, the binding energy stays high compared to the rest of the realizations (thus with different initial conditions of the binding map). After docking, amino acids that created the higher number of interactions with CS6 were Glu and Thr, and next in frequency: Lys and Asp. MD simulations show that Arg and Lys are more prone to create more ionic interactions and hydrogen bonds than the other amino acids due to their positive charge with negatively charged sulfate and carboxyl groups.

In the case of HA, a similar situation can be seen. The most stable was complex 2, i.e., composed of HSA's domains IA-IB-IIIA-IIIB. The most binding amino acids, in general, were Thr, Glu and Lys. It is very important that domains IB, IIIA, and IIIB are key domains for the HSA transport function responsible for the heme-binding site (IB), Sudlow's site II (IIIA), and thyroxine-binding site (IIIB) [45]. Comparing Table 1 to Table 2, one can notice that the HSA best binding segments differ slightly between HA and CS6 but are similar for the two first best-docked structures. In both cases, IIIA and IIIB domains clearly prevailed in creating the highest number of interactions between GAG and the protein. This is because domains IIIA and IIIB are domains that show a positive net charge on the surface that allows for binding with negatively charged GAGs. In addition, in both cases, a fragment of GAG's chain strongly interacted with an IA domain. Analyzing differences between the values of energy of binding (averaged) in both cases, HA bound to albumin about 10% stronger than CS6 (see Table S1 in Supplementary Materials).

Table 2. Binding ranks of HSA-HA complexes. The first column contains: rank after docking (averaged rank after MD simulations). The strongest connections are marked in bold letters.

HSA-HA Complex Rank	HSA Binding Sites
1(4)	IA-IB-IIIA-**IIIB**
2(1)	**IA**-IB-IIIA-IIIB
3(6)	**IA**-IB-IIIA-IIIB
4(10)	IIIA-**IIIB**
5(5)	IIB-**IIIA**-IIIB
6(8)	IA-IIIA-**IIIB**
7(2)	**IA**-IB-IIIA-IIIB
8(9)	IIIA-**IIIB**
9(7)	**IA**-IB-IIIA-IIIB
10(3)	**IIA**-IIB-IIIA

The method for computation of conformational entropy, based on a Ramachandran-type plot created for the pairs of dihedral angles (Φ,Ψ) [31,42], has been discussed in Section 2. Results of the computation have been presented in Figure 3 for the CS6 and in Figure 4 for the HA. Both of the results have been presented only for the best-bound complexes because the rest of the results had very similar characteristics. For comparison, three different realizations of YASARA simulations (thus, realizations with different initial structures) have been presented in Supplementary Materials Figures S1 and S2. The most probable angles, taken from Figures 3 and 4, have been presented in Table 3.

In the case of CS6, (Φ_{1-4}, Ψ_{1-4}) angles arranged in few clusters at ranges about: $-150° - -60°$ for Φ_{1-4} and $-180° - -60°$ for Ψ_{1-4} with the highest probability of occupancy near two more narrow angle ranges with a maximum at $(-72°, -76°)$ (cf. Figure 3, top line). There can also be seen, however, the second angle region of about: 50°–180° for Φ_{1-4} and $-180° - -50°$ for Ψ_{1-4}, also with few narrowed clusters with a high probability of occupancy. The places of the spots and their intensity differ slightly between simulations with different ionic solution. In particular, in the case of Ca^{2+}, most of the angles have been centered in one specific range around $(-104°, -76°)$, while, in the cases of Mg^{2+}

and Na^+, more than one high probability place can be seen. In the case of Mg^{2+}, the distribution of the angles is the most uniform but with, similar to the Na^+ case, a maximum at $(-68°, -76°)$.

Table 3. Most frequent (Φ_{1-4}, Ψ_{1-4}) and (Φ_{1-3}, Ψ_{1-3}) angles for the best bound complexes of HSA-GAG.

CS6						
n.o. Realization	**Na^+**		**Ca^{2+}**		**Mg^{2+}**	
2	Φ_{1-4}	Ψ_{1-4}	Φ_{1-4}	Ψ_{1-4}	Φ_{1-4}	Ψ_{1-4}
	−72°	−76°	−104°	−76°	−68°	−76°
	Φ_{1-3}	Ψ_{1-3}	Φ_{1-3}	Ψ_{1-3}	Φ_{1-3}	Ψ_{1-3}
	−76°	169°	−86°	−76°	−79°	173°
HA						
n.o. Realization	**Na^+**		**Ca^{2+}**		**Mg^{2+}**	
2	Φ_{1-4}	Ψ_{1-4}	Φ_{1-4}	Ψ_{1-4}	Φ_{1-4}	Ψ_{1-4}
	−72°	−126°	−68°	−126°	−97°	−155°
	Φ_{1-3}	Ψ_{1-3}	Φ_{1-3}	Ψ_{1-3}	Φ_{1-3}	Ψ_{1-3}
	−115°	−61°	−47°	−47°	−54°	151°

The Ca^{2+} cations are distinguished by the fact that they form many more ionic interactions with the molecules than Na^+ and Mg^{2+}. This can be the reason for the differences in the (Φ_{1-4}, Ψ_{1-4}) distribution plot. The greater probability cluster, which contains the maximum, is placed in the range of angles obtained lately in [46] for similar computer simulations of various kinds of chondroitin sulfate in water solution without any protein contribution. Thus, it can be clearly seen that the vicinity of HSA changed this crucial angle distribution, making them more disordered but still in a specific way. For proteins, similar (Φ, Ψ) angles are responsible for the formation of right-handed α-helices [42].

Despite the different chemical nature of protein and GAG molecules because peptide groups are linked at an α-carbon atom, not an oxygen like in the case of GAGs, the dihedral angles show how the chain-building units are rotated across the whole chain. Thus, the output secondary structure looks similar. In [46], there is, however, a lack of the second region on the angles, observed in our case with the positive values of Φ_{1-4} angles.

In the case of angles (Φ_{1-3},Ψ_{1-3}) for CS6, we can see quite different plots than the ones for (Φ_{1-4},Ψ_{1-4}) (see Figure 3 bottom line, and Figure S1 in Supplementary Materials). According to [46], the differences were expected because, for angles 1–3, the most probable occupancy should be in the region of −100°–−30° (Φ) and 70°–180° (Ψ). In our results, the shadow (slightly visible red color on the plot) of those angles can be seen, especially in the case of Na^+ and Mg^{2+}, where the maximum is placed within this range (cf. Table 3), near $(-76°, 169°)$. Most of the angles have had the values similar like in the case of (Φ_{1-4},Ψ_{1-4}) but more focused on regions near $(-86°, -76°)$ (maximum for Ca^{2+} case).

Conformational entropy for most cases is in the same range. However, there are noticeable differences in entropy between 1–3 and 1–4 angles. HA 1-4 angles show lower entropy i.e., are more stable than 1–3. The opposite is true for CS6. This can be explained by neighbouring groups. Carboxyl in HA and sulfur for CS6 form more stable contacts. Although acetyl group is highly reactive, it does not influence stability as dominant groups. The carboxyl group in CS6 is still weaker as compared to sulfur, which makes the contact more stable, as can be seen in Figure 5. The same behavior can be seen in Figure 6. The introduction of divalent ions increases entropy due to their destabilizing impact on protein, which is even more prominent for concentrations used in the present study.

Molecular conformational space available for HA chain in solution has been studied in [31]. The authors have searched for stable ordered forms of HA and have found many helices-type conformations (right- and left-hand side) that the HA chains prefer. Their findings based on potential energy computation for specific (Φ_{1-4},Ψ_{1-4}) and (Φ_{1-3},Ψ_{1-3}) angles are presented in Figure 2 of [31] . They obtain the location of three distinct regions with minimum potential energy surfaces. A main region consists of two wells (denoted as A–B in Figure 2 of [31] mentioned). Using the approach presented in this study, it is possible to compare the preferred dihedral angles for the HA chain, which is placed alone in the solution and in the vicinity of the HSA protein. In general, all these A-E regions (cf., Figure 2 in [31]) have been found in presented simulation results, but the intensity of these regions on the probability map varies depending on realizations (thus binding sites) and ion addition (see Figure 4 and Figure S2 in Supplementary Materials. In [31], for (Φ_{1-4},Ψ_{1-4}) angles, the main A–B region is placed in the area about -120°–-60° for Φ_{1-4} and -180°–-100° for Ψ_{1-4}, which is in accordance with results obtained in Figure 4 for the best docked HSA-HA complex. The probability of finding the angles in clusters A and B is almost equal in cases of realizations with the addition of Na^+ and Mg^{2+}, but, in the case of Ca^{2+}, B prevails over A. The maximum has been found about $(-72^\circ, -126^\circ)$ (B) and $(-97^\circ, -155^\circ)$ (A). Moreover, a few different lighter clusters have been found similar to regions C, D and E.

In Figure 4, one can see two clearly identified clusters for (Φ_{1-3},Ψ_{1-3}) angles and two smaller ones. The first of the bigger cluster, with a maximum at about $(-115^\circ, -61^\circ)$ in the case of Na^+, suits region C in [31], and the second, near the maximum $(-54^\circ, 151^\circ)$ in the case of Mg^{2+}, is placed in region A overflowing to region B. Region D, about $(50^\circ, 120^\circ)$ angles, is also present.

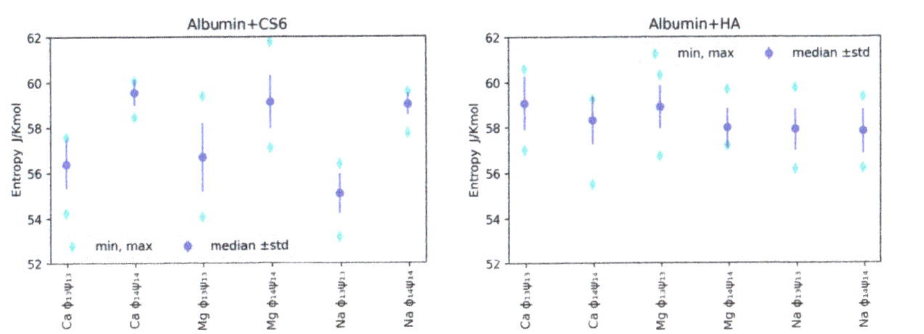

Figure 5. Median, maximal and minimal entropy (over $N = 10$ realizations) for chosen ions and angles pairs for CS6 **(left panel)** and HA (**right panel**). As we have $N = 10$ realizations, the minimal entropy can by considered as the estimate of the 10'th percentile (first quantile), and the maximal one as the estimate of the 90'th percentile (9'th quantile).

Median, minimal and maximal values (over realizations) of entropy for various ions, angles and GAGs are presented in Figure 5. Bear in mind that, as we used $N = 10$ realizations, the minimal value can be used to roughly estimate first quantile, while the maximal value to roughly estimate the 9'th quantile.

In Figure 5, one can observe that, for analysed angles, ions and GAG type, entropy was in general greater for (Φ_{1-4},Ψ_{1-4}) angles than (Φ_{1-3},Ψ_{1-3}) in the case of CS6, but it was the opposite situation for HA: the entropy was slightly greater for (Φ_{1-3},Ψ_{1-3}). As lower entropy shows lesser disorder in the system, the most stable systems were those with Na^+ ions added in both cases: HA and CS6 (cf., Figure 5). Complexes with Ca^{2+} ions usually have had slightly higher entropy. Referring to the informative interpretation of entropy, one can conclude that, in the case of CS6, the pair (Φ_{1-3},Ψ_{1-3}) carries significantly less information of the system than the pair (Φ_{1-4},Ψ_{1-4}). This is not the case for HA. The difference is related to the presence of the sulfate group in GalNAc in CS6 that is more prone to forming hydrogen bonds and ionic contacts with HSA.

Entropy values for CS6 and HA with different ions, taken separately for each of computer experiment realizations, have been presented in Figure 6.

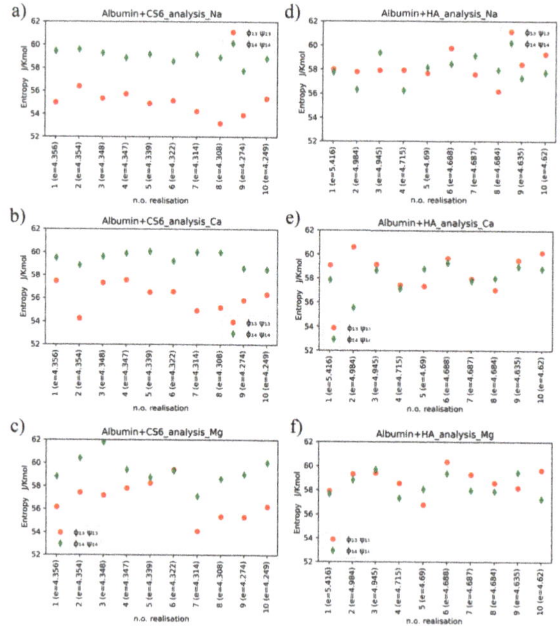

Figure 6. Conformational entropy for CS6 with (**a**) Na^+, (**b**) Ca^{2+}, and (**c**) Mg^{2+}, and HA with (**d**) Na^+, (**e**) Ca^{2+} and (**f**) Mg^{2+}.

Relatively large variations of the entropy between realizations are observed in the case of HSA-HA simulation results. Entropy varies within the range of 53–62 $\frac{J}{Kmol}$ for CS6 and within the range of 55–61 $\frac{J}{Kmol}$ for HA. Thermal noise, or some non-equilibrium processes, can have some effect on these variations, but also the different binding sites of the two molecules (that varies between realizations) are of big importance for the entropy behaviour. The variations of entropy values may also coincide with rather high estimation error of this value.

The hypothesis that entropy value is tied to the value of the binding energy between the protein and the GAG is not supported by our simulation results. In more detail, the smallest entropy value, roughly 53 $\frac{J}{Kmol}$, has been obtained for realization number 8 of the HSA-CS6 complex with Na^+. This was the case, where only IA and IB domains of HSA were bound to CS6; thus, the protein did not affect the conformation of the GAG's chain by deformation of the (Φ,Ψ) angles much. The highest entropy value has been reported for case 3 of the HSA-CS6 complex with Mg^{2+}. Here, the binding site was very similar to the one with greater entropy (number 8): IA and IIA. Entropy records for realization number 2 of the HSA+HA complex with addition of Ca^{2+} are very interesting. In this realization, there were a huge difference between entropy for (Φ_{1-4},Ψ_{1-4}) angles and (Φ_{1-3},Ψ_{1-3}) angles. In this case, the HSA and HA molecules were best bound after MD simulation from all the realizations.

4. Discussion

A common pattern on the maps presented in Figure 4 and the ones presented in Figure 2 in [31] indicates that the proximity of the HSA protein does not prevent the HA chain from taking the shape of a helix (cf., Figure 5 in [31]). In the case of CS6, positions of preferred (Φ,Ψ) angles regions have been recently reported in [46]. The location of the angles presented in this study for HSA-CS6 complexes overlaps with those for the HA chain in [31]. Therefore, it can be concluded that CS6 keeps its α-helix structure.

However, distributions of the (Φ,Ψ) angles (Figures 3 and 4) show that GAGs can show helix structures and the random coil conformations at various ratios. Bear in mind that the frequency distributions of the (Φ,Ψ) angles are created from all NG units over the whole time; thus, we cannot say anything about how the conformations of the chains evolved. Instead, we see the static characteristic maps, which can tell us about the stability of the conformations in the simulation process.

We have demonstrated that conformational entropy is a parameter that enables us to characterize the structure of GAGs in interaction with HSA globally. We can see that it is slightly dependent on the pair of angles but rather for the CS6 case. In the case of CS6, the lower entropy value of the (Φ_{1-3},Ψ_{1-3}) angles compared to the entropy value of (Φ_{1-4},Ψ_{1-4}) angles indicate that mathematical features of the histograms for the angle pair 1-4 relative to the angle pair 1–3 must be more uniform. We understand this idea that (Φ_{1-4},Ψ_{1-4}) angles during simulations appear more uniform in the space of its value than angles (Φ_{1-3},Ψ_{1-3}) in the space of its value. This property is not visible for the angles in the case of HA.

Numerically derived histograms have various numbers of maxima, but these local maxima may be wider or narrower, which is essential in the entropy calculated here. Therefore, our work builds upon [46], which focuses on a sole CS6 and shows that a smaller number of regions are preferred to occupy by (Φ,Ψ) angles than our work that shows a larger number where HSA was added to the solution.

As the entropy is one of the features of above-mentioned histograms, we may expect other features of these histograms to be also informative about particular GAG, ion, or monomer. Hence, further analysis concerning pattern recognition machine learning techniques like SVM (Support Vector Machine) [47] and its modification for colored image processing [48] can be performed to extend the research. Another approach to this problem may be the colored image segmentation by a random walk [49], Sub-Markov random walk [50] or the Hurst Exponent image processing in [51,52]. Alternatively, one can analyze angles as multivariate series and process them with dedicated tools such as higher-order multivariate cumulants, see [53]. The goal of deep analysis mentioned above would be to search for more sensitive features and analyze whether entropy is a stable feature compared to others. If such features can be determined, their utility may appear in analytical, chemical medical applications. The secondary GAGs structure resembles flat bands transformed into a helix or twisted into a sheet originating from intermolecular hydrogen bonds. In diluted GAG solutions, the macromolecules have semi-rigid coiled chains and could form helix bands and even helical rings. Due to formation of a rigid helix, the macromolecules of GAG attract a great quantity of water and organize the broad domains of the tertiary polymer structure [54]. Binding to HSA reduces degrees of freedom of polymer and thus mechanical properties. On the other hand, HA and CS6 at higher concentrations, external force or other factors can change HSA's tertiary structure, forming material of different properties.

5. Conclusions

Macromolecular complexes are building blocks in the functioning of physiological processes. When optimal conditions are fulfilled for given pairs, the system can function efficiently. Protein–ligand interactions are critical to optimal biochemical, biological, or biophysical results. Often, a given complex can serve several functions, as in the case presented in this study, where HA-HSA complexes decrease friction and can be used in drug delivery systems. Our results show how crucial components of synovial fluid interplay with each other at equilibrium. We have shown that HA and CS6 can form stable complexes with HSA.

Moreover, the binding sites for both molecules overlap, which indicates that they both can induce a similar effect on HSA while functioning. The molecular mass used in this study is one limitation of the presented results, as interactions (and mechanical and biological properties) between GAGs and proteins strongly depend on their molecular mass and concentration. This fact emerges from the chemistry of polymers of interest.

GAGs chain's amount of expansion is enormous for a semi-flexible polymer. The polymer configuration is constantly in a state of motion and change. However, the water increases the effective size of each hyaluronic acid because of its hydrophilic nature. The mass increase results in the average density decrease because the increase in mass is slower than in the volume. Thus, GAGs chains with a high molecular weight (more than 1,000 kDa) occupy a substantial volume. However, adsorption at HSA strongly influences the local mechanical properties of GAGs resulting in efficient lubrication.

Supplementary Materials: The following supporting information can be downloaded at: https://www.mdpi.com/article/10.3390/e24060811/s1, Figure S1: Normalized histograms for different realizations for HSA-CS6 complexes.; Figure S2: Normalized histograms for different realizations for HSA-HA complexes.; Table S1: Binding energies in kcal/mol for 10 realizations of HSA-CS6 and HSA-HA complexes after docking.

Author Contributions: Conceptualization, P.S. and P.B.; methodology, P.B., N.K., K.D. and P.S.; software, P.S., K.D. and N.K.; validation, N.K., P.B., P.W. and B.M.; formal analysis, K.D., P.S., P.B., N.K. and P.W.; investigation, P.B., N.K., K.D., P.S., P.W. and B.M.; resources, P.S., P.B., P.W., N.K., K.D. and B.M.; data curation, P.B., N.K. and P.S.; writing—original draft preparation, P.B., P.W., N.K. and K.D.; writing—review and editing, P.B., N.K. and K.D.; visualization, P.B., N.K. and P.S.; supervision, K.D. and P.B.; project administration, P.B.; funding acquisition, B.M. All authors have read and agreed to the published version of the manuscript.

Funding: This research received no external funding.

Institutional Review Board Statement: Not applicable.

Informed Consent Statement: Not applicable.

Data Availability Statement: Data and processing code is available in the public repository at [41], https://github.com/iitis/polymer_entropy accessed on 7 June 2022.

Acknowledgments: The work is supported by BN-WTiICh-11/2022 of the Bydgoszcz University of Science and Technology. Calculations were carried out at the Academic Computer Centre in Gdańsk.

Conflicts of Interest: The authors declare no conflict of interest.

Abbreviations

The following abbreviations are used in this manuscript:

HSA	human serum albumin
HA	hyaluronic acid
CS6	chondroitin 6 sulfate
GAG	glycosaminoglycan
MD	molecular dynamics
PDB	protein data bank

References

1. Klein, J.; Raviv, U.; Perkin, S.; Kampf, N.; Chai, L.; Giasson, S. Fluidity of water and of hydrated ions confined between solid surfaces to molecularly thin films. *J. Phys. Condens. Matter* **2004**, *16*, S5437–S5448. [CrossRef]
2. Gadomski, A.; Pawlak, Z.; Oloyede, A. Directed ion transport as virtual cause of some facilitated friction–lubrication mechanism prevailing in articular cartilage: A hypothesis. *Tribol. Lett.* **2008**, *30*, 83–90. [CrossRef]
3. Gadomski, A.; Bełdowski, P.; Rubí, J.M.; Urbaniak, W.; Augé, W.K.; Santamaría-Holek, I.; Pawlak, Z. Some conceptual thoughts toward nanoscale oriented friction in a model of articular cartilage. *Math. Biosci.* **2013**, *244*, 188–200. [CrossRef]
4. Dėdinaitė, A.; Claesson, P.M. Synergies in lubrication. *Phys. Chem. Chem. Phys.* **2017**, *19*, 23677–23689. [CrossRef]
5. Raj, A.; Wang, M.; Zander, T.; Wieland, D.F.; Liu, X.; An, J.; Garamus, V.M.; Willumeit-Römer, R.; Fielden, M.; Claesson, P.M.; et al. Lubrication synergy: Mixture of hyaluronan and dipalmitoylphosphatidylcholine (DPPC) vesicles. *J. Colloid Interface Sci.* **2017**, *488*, 225–233. [CrossRef]
6. Klein, J. Molecular mechanisms of synovial joint lubrication. *Proc. Inst. Mech. Eng. Part J J. Eng. Tribol.* **2006**, *220*, 691–710. [CrossRef]
7. Liu, C.; Wang, M.; An, J.; Thormann, E.; Dėdinaitė, A. Hyaluronan and phospholipids in boundary lubrication. *Soft Matter* **2012**, *8*, 10241–10244. [CrossRef]

8. Siódmiak, J.; Bełdowski, P.; Augé, W.; Ledziński, D.; Śmigiel, S.; Gadomski, A. Molecular dynamic analysis of hyaluronic acid and phospholipid interaction in tribological surgical adjuvant design for osteoarthritis. *Molecules* **2017**, *22*, 1436. [CrossRef] [PubMed]
9. Ghosh, S.; Choudhury, D.; Das, N.S.; Pingguan-Murphy, B. Tribological role of synovial fluid compositions on artificial joints — A systematic review of the last 10 years. *Lubr. Sci.* **2014**, *26*, 387–410. [CrossRef]
10. Boldt, J. Use of albumin: An update. *BJA Br. J. Anaesth.* **2010**, *104*, 276–284. [CrossRef] [PubMed]
11. Moman, R.N.; Gupta, N.; Varacallo, M. *Physiology Albumin*; StatPearls Publishing: Treasure Island, FL, USA, 2021.
12. Ghosh, S.; Choudhury, D.; Pingguan-Murphy, B. Lubricating ability of albumin and globulin on artificial joint implants: a tribological perspective. *Int. J. Surf. Sci. Eng.* **2016**, *10*, 193–206. [CrossRef]
13. Nečas, D.; Sadecká, K.; Vrbka, M.; Galandáková, A.; Wimmer, M.; Gallo, J.; Hartl, M. The effect of albumin and γ-globulin on synovial fluid lubrication: Implication for knee joint replacements. *J. Mech. Behav. Biomed. Mater.* **2021**, *113*, 104117. [CrossRef]
14. Kubiak-Ossowska, K.; Jachimska, B.; Mulheran, P.A. How Negatively Charged Proteins Adsorb to Negatively Charged Surfaces: A Molecular Dynamics Study of BSA Adsorption on Silica. *J. Phys. Chem. B* **2016**, *120*, 10463–10468. [CrossRef] [PubMed]
15. Bełdowski, P.; Przybyłek, M.; Raczyński, P.; Dedinaite, A.; Górny, K.; Wieland, F.; Dendzik, Z.; Sionkowska, A.; Claesson, P.M. Albumin–Hyaluronan Interactions: Influence of Ionic Composition Probed by Molecular Dynamics. *Int. J. Mol. Sci.* **2021**, *22*, 12360. [CrossRef] [PubMed]
16. Guizado, T.R.C. Analysis of the structure and dynamics of human serum albumin. *J. Mol. Model.* **2014**, *20*, 2450. [CrossRef] [PubMed]
17. Shi, D.; Sheng, A.; Chi, L. Glycosaminoglycan-Protein Interactions and Their Roles in Human Disease. *Front. Mol. Biosci.* **2021**, *8*, 639666. [CrossRef]
18. Qiu, H.; Jin, L.; Chen, J.; Shi, M.; Shi, F.; Wang, M.; Li, D.; Xu, X.S.; Su, X.; Yin, X.; et al. Comprehensive Glycomic Analysis Reveals That Human Serum Albumin Glycation Specifically Affects the Pharmacokinetics and Efficacy of Different Anticoagulant Drugs in Diabetes. *Diabetes* **2020**, *69*, 760–770. [CrossRef]
19. Krieger, E.; Vriend, G. YASARA View—Molecular graphics for all devices–From smartphones to workstations. *Bioinformatics* **2014**, *30*, 2981–2982. [CrossRef]
20. Gandhi, N.S.; Mancera, R.L. The Structure of Glycosaminoglycans and their Interactions with Proteins. *Chem. Biol. Drug Des.* **2008**, *72*, 455–482. [CrossRef]
21. Bełdowski, P.; Yuvan, S.; Dėdinaitė, A.; Claesson, P.M.; Pöschel, T. Interactions of a short hyaluronan chain with a phospholipid membrane. *Colloids Surf. B Biointerfaces* **2019**, *184*, 110539. [CrossRef]
22. Ben-Naim, A. *Molecular Theory of Water and Aqueous Solutions*; World Scientific Publishing Company: Singapore, 2011.
23. Baruah, A.; Rani, P.; Biswas, P. Conformational entropy of intrinsically disordered proteins from amino acid triads. *Sci. Rep.* **2015**, *5*, 11740. [CrossRef] [PubMed]
24. Kruszewska, N.; Bełdowski, P.; Domino, K.; Lambert, K.D. Investigating conformation changes and network formation of mucin in joints functioning in human locomotion. In *Multiscale (Loco)motion— Toward Its Active-Matter Addressing Physical Principles*; Gadomski, A., Ed.; UTP Publishing Department: Bydgoszcz, Poland, 2019; pp. 121–138.
25. Sapienza, P.J.; Lee, A.L. Using NMR to study fast dynamics in proteins: methods and applications. *Curr. Opin. Pharmacol.* **2010**, *10*, 723–730. [CrossRef] [PubMed]
26. Thompson, J.B.; Hansma, H.G.; Hansma, P.K.; Plaxco, K.W. The backbone conformational entropy of protein folding: Experimental measures from Atomic Force Microscopy. *J. Mol. Biol.* **2002**, *322*, 645–652. [CrossRef]
27. Fitter, J. A Measure of Conformational Entropy Change during Thermal Protein Unfolding Using Neutron Spectroscopy. *Biophys. J.* **2003**, *84*, 3924–3930. [CrossRef]
28. Meirovitch, H.; Cheluvaraja, S.; White, R. Methods for calculating the entropy and free energy and their application to problems involving protein flexibility and ligand binding. *Curr. Protein Pept. Sci.* **2009**, *10*, 229–243. [CrossRef]
29. Bhattacharjee, N.; Biswas, P. Are ambivalent α-helices entropically driven? *Protein Eng. Des. Sel. PEDS* **2012**, *25*, 73–79. [CrossRef]
30. Baxa, M.C.; Haddadian, E.J.; Jumper, J.M.; Freed, K.F.; Sosnick, T.R. Loss of conformational entropy in protein folding calculated using realistic ensembles and its implications for NMR-based calculations. *Proc. Natl. Acad. Sci. USA* **2014**, *111*, 15396–15401. [CrossRef]
31. Haxaire, K.; Braccini, I.; Milas, M.; Rinaudo, M.; Pérez, S. Conformational behavior of hyaluronan in relation to its physical properties as probed by molecular modeling. *Glycobiology* **2000**, *10*, 587–594. [CrossRef]
32. Trott, O.; Olson, A.J. AutoDock Vina: Improving the speed and accuracy of docking with a new scoring function, efficient optimization, and multithreading. *J. Comput. Chem.* **2010**, *31*, 455–461. [CrossRef]
33. Duan, Y.; Wu, C.; Chowdhury, S.; Lee, M.C.; Xiong, G.; Zhang, W.; Yang, R.; Cieplak, P.; Luo, R.; Lee, T.; et al. A point-charge force field for molecular mechanics simulations of proteins based on condensed-phase quantum mechanical calculations. *J. Comput. Chem.* **2003**, *24*, 1999–2012. [CrossRef]
34. Maier, J.A.; Martinez, C.; Kasavajhala, K.; Wickstrom, L.; Hauser, K.E.; Simmerling, C. ff14SB: Improving the Accuracy of Protein Side Chain and Backbone Parameters from ff99SB. *J. Chem. Theory Comput.* **2015**, *11*, 3696–3713. [CrossRef] [PubMed]
35. Kirschner, K.N.; Yongye, A.B.; Tschampel, S.M.; González-Outeiriño, J.; Daniels, C.R.; Foley, B.L.; Woods, R.J. GLYCAM06: A generalizable biomolecular force field. Carbohydrates. *J. Comput. Chem.* **2008**, *29*, 622–655. . [CrossRef] [PubMed]
36. Krieger, E.; Dunbrack, R.L.; Hooft, R.W.; Krieger, B. Assignment of protonation states in proteins and ligands: Combining PKA prediction with hydrogen bonding network optimization. *Methods Mol. Biol.* **2012**, *819*, 405–421. [CrossRef] [PubMed]

37. Krieger, E.; Vriend, G. New ways to boost molecular dynamics simulations. *J. Comput. Chem.* **2015**, *36*, 996–1007. [CrossRef]
38. Hornak, V.; Abel, R.; Okur, A.; Strockbine, B.; Roitberg, A.E.; Simmerling, C. Comparison of multiple Amber force fields and development of improved protein backbone parameters. *Proteins Struct.* **2006**, *65*, 712–725. [CrossRef]
39. Essmann, U.; Perera, L.E.; Berkowitz, M.L.; Darden, T.A.; Lee, H.C.; Pedersen, L.G. A smooth particle mesh Ewald method. *J. Chem. Phys.* **1995**, *103*, 8577–8593. [CrossRef]
40. Berendsen, H.J.C.; Postma, J.P.M.; van Gunsteren, W.F.; Dinola, A.; Haak, J.R. Molecular dynamics with coupling to an external bath. *J. Chem. Phys.* **1984**, *81*, 3684–3690. [CrossRef]
41. Sionkowski, P.; Domino, K.; Kruszewska, N. Polymer_Entropy. 2022. Available online: https://github.com/iitis/polymer_entropy (accessed on 9 June 2022)
42. Ramachandran, G.; Ramakrishnan, C.; Sasisekharan, V. Stereochemistry of polypeptide chain configurations. *J. Mol. Biol.* **1963**, *7*, 95–99. [CrossRef]
43. Shannon, C.E. A mathematical theory of communication. *ACM SIGMOBILE Mob. Comput. Commun. Rev.* **2001**, *5*, 3–55. [CrossRef]
44. Weber, P.; Bełdowski, P.; Domino, K.; Ledziński, D.; Gadomski, A. Changes of Conformation in Albumin with Temperature by Molecular Dynamics Simulations. *Entropy* **2020**, *22*, 405. [CrossRef]
45. Fasano, M.; Curry, S.; Terreno, E.; Galliano, M.; Fanali, G.; Narciso, P.; Notari, S.; Ascenzi, P. The extraordinary ligand binding properties of human serum albumin. *IUBMB Life* **2005**, *57*, 787–796. [CrossRef] [PubMed]
46. Nagarajan, B.; Sankaranarayanan, N.V.; Desai, U.R. In-depth molecular dynamics study of all possible chondroitin sulfate disaccharides reveals key insight into structural heterogeneity and dynamism. *Biomolecules* **2022**, *12*, 77. [CrossRef] [PubMed]
47. Burges, C.J. A tutorial on support vector machines for pattern recognition. *Data Min. Knowl. Discov.* **1998**, *2*, 121–167. [CrossRef]
48. Khemchandani, R.; Saigal, P. Color image classification and retrieval through ternary decision structure based multi-category TWSVM. *Neurocomputing* **2015**, *165*, 444–455. [CrossRef]
49. Grady, L. Random walks for image segmentation. *IEEE Trans. Pattern Anal. Mach. Intell.* **2006**, *28*, 1768–1783. [CrossRef]
50. Dong, X.; Shen, J.; Shao, L.; Van Gool, L. Sub-Markov random walk for image segmentation. *IEEE Trans. Image Process.* **2015**, *25*, 516–527. [CrossRef]
51. Blachowicz, T.; Ehrmann, A.; Domino, K. Statistical analysis of digital images of periodic fibrous structures using generalized Hurst exponent distributions. *Phys. A Stat. Mech. Appl.* **2016**, *452*, 167–177. [CrossRef]
52. Blachowicz, T.; Domino, K.; Koruszowic, M.; Grzybowski, J.; Böhm, T.; Ehrmann, A. Statistical analysis of nanofiber mat AFM images by Gray-scale-resolved Hurst exponent distributions. *Appl. Sci.* **2021**, *11*, 2436. [CrossRef]
53. Domino, K. Multivariate cumulants in outlier detection for financial data analysis. *Phys. A Stat. Mech. Appl.* **2020**, *558*, 124995. [CrossRef]
54. Snetkov, P.; Zakharova, K.; Morozkina, S.; Olekhnovich, R.; Uspenskaya, M. Hyaluronic Acid: The Influence of Molecular Weight on Structural, Physical, Physico-Chemical, and Degradable Properties of Biopolymer. *Polymers* **2020**, *12*, 1800. [CrossRef]

entropy

MDPI

Article

Does Social Distancing Matter for Infectious Disease Propagation? An SEIR Model and Gompertz Law Based Cellular Automaton

Szymon Biernacki and Krzysztof Malarz *

Faculty of Physics and Applied Computer Science, AGH University of Science and Technology, al. Mickiewicza 30, 30-059 Kraków, Poland; szy111mon@wp.pl
* Correspondence: malarz@agh.edu.pl

Abstract: In this paper, we present stochastic synchronous cellular automaton defined on a square lattice. The automaton rules are based on the SEIR (susceptible → exposed → infected → recovered) model with probabilistic parameters gathered from real-world data on human mortality and the characteristics of the SARS-CoV-2 disease. With computer simulations, we show the influence of the radius of the neighborhood on the number of infected and deceased agents in the artificial population. The increase in the radius of the neighborhood favors the spread of the pandemic. However, for a large range of interactions of exposed agents (who neither have symptoms of the disease nor have been diagnosed by appropriate tests), even isolation of infected agents cannot prevent successful disease propagation. This supports aggressive testing against disease as one of the useful strategies to prevent large peaks of infection in the spread of SARS-CoV-2-like diseases.

Keywords: epidemy; compartmental models; computer simulation; SARS-CoV-2-like disease spreading

Citation: Biernacki, S.; Malarz, K. Does Social Distancing Matter for Infectious Disease Propagation? An SEIR Model and Gompertz Law Based Cellular Automaton. *Entropy* **2022**, *24*, 832. https://doi.org/10.3390/e24060832

Academic Editor: Adam Gadomski

Received: 25 March 2022
Accepted: 11 June 2022
Published: 15 June 2022

1. Introduction

Currently, the death rate of SARS-CoV-2 [1] in the whole world reached around 1.2% of the infected population and more than 5×10^8 cases have been confirmed [2,3], (tab 'Closed Cases') . Thus, one should not be surprised of the publication rash in this subject giving (i) theoretical bases of SARS-CoV-2 spreading, (ii) practical tips on preventing plagues or even (iii) clinical case studies making it easier to recognize and to treat cases of the disease. The Web of Science database reveals over 80,000 and over 110,000 papers related to this topic registered in 2020 and in January–November 2021, respectively, in contrast to only 19 papers in 2019. Among them, only several [4–20] are based on the cellular automata technique [21–24].

The likely reason for this moderate interest in using this technique to simulate the spread of the COVID-19 pandemic is the large degree of simplification of 'rules of the game' in cellular automata. To fill this gap, in this work, we propose a cellular automaton based on a compartmental model, the parameters of which were adjusted to the realistic probabilities of the transitions between the states of the automaton. Let us note that modeling the spread of the pandemic is also possible with other models (see Refs. [25–28] for mini-reviews) including, for instance, those based on the percolation theory [29].

The history of the application of compartmental models to the mathematical modeling of infectious diseases dates to the first half of the 20th century and works of Ross [30], Ross and Hudson [31,32], Kermack and McKendrick [33,34] and Kendall [35]; see, for instance, Ref. [36] for an excellent review. In the compartmental model, the population is divided into several (usually labeled) compartments so that the agent only remains in one of them and the sequences of transitions between compartments (label changes) are defined. For instance, in the classical SIR model, agents change their states subsequently from *susceptible* ($\mathcal{S}$) via *infected* ($\mathcal{I}$) to *recovered* ($\mathcal{R}$) [33,34,37]. Infected agents can transmit the disease to their susceptible neighbors ($\mathcal{S} \to \mathcal{I}$) with a given probability p_1. The infected agent may recover ($\mathcal{I} \to \mathcal{R}$) with probability p_2. After recovering, the agents are immune and they can no longer be infected with the disease. These rules may be described by a set of differential equations:

$$\begin{aligned} \frac{dn_{\mathcal{S}}}{dt} &= -\langle k \rangle p_1 n_{\mathcal{S}} n_{\mathcal{I}}, \\ \frac{dn_{\mathcal{I}}}{dt} &= \langle k \rangle p_1 n_{\mathcal{S}} n_{\mathcal{I}} - p_2 n_{\mathcal{I}}, \\ \frac{dn_{\mathcal{R}}}{dt} &= p_2 n_{\mathcal{I}}, \end{aligned} \tag{1}$$

where $\langle k \rangle$ is the mean number of agents' contacts in the neighborhood and $n_{\mathcal{S}}$, $n_{\mathcal{I}}$, $n_{\mathcal{R}}$ represent the fraction of susceptible, infected, and recovered agents, respectively. Typically, the initial condition for Equation (1) is:

$$\begin{aligned} n_{\mathcal{S}}(t=0) &= 1 - n_0, \\ n_{\mathcal{I}}(t=0) &= n_0, \\ n_{\mathcal{R}}(t=0) &= 0, \end{aligned} \tag{2}$$

where n_0 is the initial fraction of infected agents (fraction of 'Patients Zero').

The transition rates between states (i.e., probabilities p_1 and p_2) may be chosen arbitrarily or they may correspond to the reciprocal of agents' residence times in selected states. In the latter case, residence time may be estimated by clinical observations [38,39]. The probability p_1 describes the speed of disease propagation (infecting rate) while the value of p_2 is responsible for the frequency of getting better (recovering rate). In this approximation, the dynamics of the infectious class depends on the reproduction ratio:

$$R_0 = \frac{\langle k \rangle p_1}{p_2}. \tag{3}$$

The case of $R_0 = 1$ separates the phase when the disease dies out and the phase when the disease spreads among the members of the population.

Equation (1) describe a mean-field evolution, which simulates a situation in which all agents interact directly with each other. In low-dimensional spatial networks, the mean-field dynamics (1) is modified by diffusive mechanisms [40]. In a realistic situation, the diffusive mode of pandemic spreading is mixed with the mean-field dynamics, corresponding to nonlocal transmissions resulting from the mobility of agents [41].

We use the SEIR model [42,43], upon extending the SIR model, where an additional compartment (labeled $\mathcal{E}$) is available and it corresponds to agents in the *exposed* state. The exposed agents are infected but unaware of it—they neither have symptoms of the disease nor have been diagnosed by appropriate tests. This additional state requires splitting the transition rate p_1 into $p_{\mathcal{E}}$ and $p_{\mathcal{I}}$ corresponding to transition rates (probabilities) $\mathcal{S} \to \mathcal{E}$ after contact with the exposed agent in state $\mathcal{E}$ and $\mathcal{S} \to \mathcal{E}$ after contact with the infected agent in state $\mathcal{I}$, respectively. We would like to emphasize that both exposed (in state $\mathcal{E}$) and infected (in state $\mathcal{I}$) agents *may transmit* disease.

According to Equation (1), after recovering, the convalescent in state $\mathcal{R}$ lives forever, which seems contradictory to the observations of the real world. Although the Bible Book of Genesis (5:5–27; 9:29) mentions seven men who lived over 900 years, in modern society—thanks to public health systems (and sometimes in spite of them)—contemporary living lengths beyond one hundred years are rather rare. Mortality tables [44] show some correlations between probability of death and age [45]. This observation was first published by Gompertz in 1825 [46]. According to Gompertz's law, mortality f increases exponentially with the age a of the individual as:

$$f(a) \propto \exp(b(a+c)), \tag{4}$$

where b and c are constants. Moreover, as we mentioned in the first sentence of the Introduction, people can also die earlier than Gompertz's law implies. For example, an epidemic of fatal diseases increases the mortality rate. To take care of these factors in modeling disease propagation, we consider removing agents from the population. This happens with agents' age-dependent probabilities f_G and f_C for healthy and ill people, respectively. The removed individual is immediately replaced with a newly born baby.

In this paper, we propose a cellular automaton based simultaneously on SEIR model of disease propagation and Gompertz's law of mortality. In Section 2, the cellular automaton, its rules and the available site neighborhoods are presented. Section 3 is devoted to the presentation of the results of simulations based on the proposed cellular automaton. Discussion of the obtained results (Section 4) and conclusions (Section 5) end the paper. We note [47,48], where age-structured populations were also studied with SEIR-based and multicompartments models, respectively.

2. Model

We use the cellular automata technique [21–24] to model disease propagation. The cellular automata technique is based on several assumptions, including:

- Discrete (geometrical) space (i.e., regular lattice) and time;
- Discrete and finite set of available states of the single lattice's site;
- Local rule $\mathcal{F}$ of synchronous site states update.

The rule $\mathcal{F}$ defines the state s_i of the site i at time $(t+1)$ based on this site's state s_i at time t and the state of the sites in the i-th site's neighborhood $\mathcal{N}$

$$s_i(t+1) = \mathcal{F}\Big(s_i(t); \mathcal{N}(s_i(t))\Big).$$

We adopt the SEIR model for the simulation of disease spreading by probabilistic synchronous cellular automata on a square lattice with various neighborhoods $\mathcal{N}$. Every agent may be in one of four available states: susceptible ($\mathcal{S}$), exposed ($\mathcal{E}$), infected ($\mathcal{I}$) or recovered ($\mathcal{R}$). The agents in $\mathcal{E}$ and $\mathcal{I}$ are characterized by different values of the probability of infection ($p_{\mathcal{E}}$ and $p_{\mathcal{I}}$) and different ranges of interactions (radii of neighborhoods $r_{\mathcal{E}}$ and $r_{\mathcal{I}}$). The considered neighborhoods (and their radii) are presented in Figure 1.

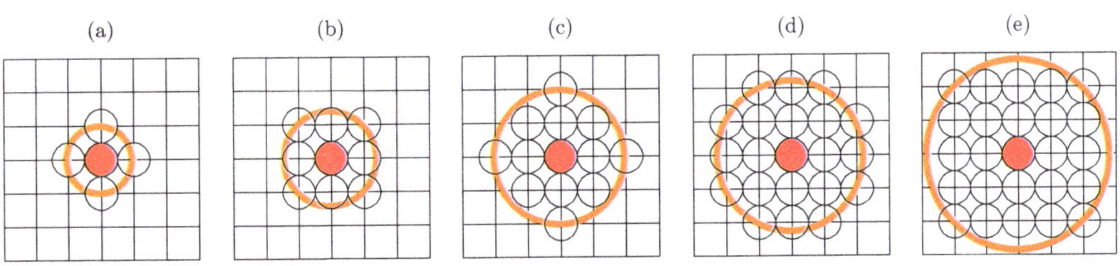

Figure 1. Sites in various neighborhoods $\mathcal{N}$ on a square lattice. (**a**) Von Neumann's neighborhood ($r = 1, z = 4$); (**b**) Moore's neighborhood ($r = \sqrt{2}, z = 8$); (**c**) neighborhood with sites up to the third coordination zone ($r = 2, z = 12$); (**d**) neighborhood with sites up to the fourth coordination zone ($r = \sqrt{5}, z = 20$); (**e**) neighborhood with sites up to the fifth coordination zone ($r = 2\sqrt{2}, z = 24$).

Initially (at $t = 0$), every agent is in the $\mathcal{S}$ state, their age a is set randomly from a normal distribution with the mean value $\langle a \rangle = 50 \times 365$ days and dispersion $\sigma^2 = 25 \times 365$ days. 'Patient Zero' in the $\mathcal{E}$ state is placed randomly at a single site of a square lattice with $L^2 = 100^2$ nodes.

Every time step (which corresponds literally to a single day in the real-world) the lattice is scanned in typewriter order to check the possible agent state evolution:

- The susceptible agent may be infected ($\mathcal{S} \to \mathcal{E}$) by each agent in state $\mathcal{E}$ or $\mathcal{I}$ present in his/her neighborhood $\mathcal{N}$ with radius $r_{\mathcal{E}}$ or $r_{\mathcal{I}} \le r_{\mathcal{E}}$, respectively. We set $r_{\mathcal{I}} \le r_{\mathcal{E}}$

as we assume that infected (and aware of the disease) agents are more responsible than exposed (unaware of the disease) ones. The latter inequality comes from our assumption that the exposed agent is not careful enough in undertaking contacts with his/her neighbours while the infected agent is serious-minded and realizing the hazard of possible disease propagation and thus he/she avoids these contacts at least on the level assigned to exposed agents. The number and position of available neighbours who may infect the considered susceptible agent depend on the value of radius $r_{\mathcal{E}}$ and/or $r_{\mathcal{I}}$ as presented in Figure 1. The infection of the susceptible agents occurs with probability $p_{\mathcal{E}}$ (after contacting with agent in state $\mathcal{E}$) or $p_{\mathcal{I}}$ (after contacting with agent in state $\mathcal{I}$), respectively.

- The incubation (i.e., the appearance of disease symptoms) takes $\tau_{\mathcal{E}}$ days—every agent in $\mathcal{E}$ state is converted to infected state ($\mathcal{E} \to \mathcal{I}$) with probability $1/\tau_{\mathcal{E}}$. The exposed agent may die with age-dependent probability $f_C(a)$. In such a case, he/she is replaced ($\mathcal{E} \to \mathcal{S}$) with newly born agent ($a = 0$).
- The disease lasts for $\tau_{\mathcal{I}}$ days. The ill agent (in state $\mathcal{I}$) may either die (and be replaced by a newly born child $\mathcal{I} \to \mathcal{S}$) with age specific probability $f_C(a)$ or he/she can recover (and gain resistance to disease $\mathcal{I} \to \mathcal{R}$) with probability $1/\tau_{\mathcal{I}}$.
- A healthy agent ($\mathcal{S}$ or $\mathcal{R}$) may die with a chance given by age dependent probability $f_G(a)$. In such a case, it is replaced with a newly born susceptible baby (in state $\mathcal{S}$ and in age of $a = 0$).

The agents' state modifications are applied synchronously to all sites. A single time step (from t to $t+1$) of the system evolution described above is presented in Algorithm 1. The L^2-long vector variables `tmp_pop[]` and `pop[]` represent the current population (at time t) and the population in the next time step ($t+1$), respectively. The presence of two such variables in a model implementation is caused by the synchronicity of cellular automaton. The i-th element of these vectors keeps information on the state (either $\mathcal{S}, \mathcal{E}, \mathcal{I}$ or $\mathcal{R}$) of the i-th agent in the population. The age a (measured in days) for the i-th agent is kept in the L^2-long vector `age[]` and its i-th element is either incremented (line 4 of Algorithm 1) or reset (lines 28 and 35 of Algorithm 1) in the case of removal. The `random()` function returns a real pseudo-random number uniformly distributed in $[0,1)$. The function $||i,j||$ measures the Euclidean distance between agents i and j. The age-dependent daily death probability functions $f_C(\cdot)$ and $f_G(\cdot)$—based on real-world data—are defined in the next paragraph in Equations (5) and (6). The numbers $n_{\mathcal{S}}, n_{\mathcal{E}}, n_{\mathcal{I}}$ and $n_{\mathcal{R}}$ of agents in various states must be initialized at $t = 0$ based on initial conditions. The cumulative number $n_{\mathcal{D}}$ of deceased agents—earlier either in the exposed ($\mathcal{E}$) or in the infected ($\mathcal{I}$) state—is incremented in line 39 of Algorithm 1. The numbers $n_{\mathcal{S}}, n_{\mathcal{E}}, n_{\mathcal{I}}, n_{\mathcal{R}}$ and $n_{\mathcal{D}}$ are normalized to the total number L^2 of agents in the system before their return.

Based on American data on annual death probability [44] and assuming 365 days a year, we predict the daily death probability as:

$$f_G(a) = 184 \cdot 10^{-10} \exp(0.00023(a + 40259)), \tag{5}$$

where a is the agent's age expressed in days. The data follow Gompertz's exponential law of mortality [46,49].

Using the same trick, we estimate the probability of daily death for infected people of age a (expressed in days) as:

$$f_C(a) = \begin{cases} 5 \cdot 10^{-5} & \iff a \le 30\,y, \\ 2 \cdot 10^{-6} \exp(0.0003a) & \iff a > 30\,y. \end{cases} \tag{6}$$

These probabilities are calculated as the chance of death during SARS-CoV-2 infection (based on Polish statistics [50]) divided by $(\tau_{\mathcal{E}} + \tau_{\mathcal{I}})$. We assume that infection lasts $\tau_{\mathcal{I}} \approx 14$ days and that an incubation process takes $\tau_{\mathcal{E}} \approx 5$ days [39]. Exponential fits (Equations (5) and (6)) to real-world data are presented in Figure 2.

Algorithm 1 Single time step in automaton

- $t \leftarrow t + 1$
- tmp_pop $\leftarrow$ pop
- **for all** $i \in$ pop **do**
  - age[i] $\leftarrow$ age[i]+1
  - **if** tmp_pop[i] = $\mathcal{S}$ **then**
    - **for all** $j \in$ pop, $j \neq i$ **do**
      - **if** tmp_pop[j] = $\mathcal{I} \wedge \|i, j\| \leq r_{\mathcal{I}} \wedge$ random() $\leq p_{\mathcal{I}}$ **then**
        - pop[i] $\leftarrow \mathcal{E}$ ▷ *$\mathcal{S} \to \mathcal{E}$ after contacting with $\mathcal{I}$*
        - $n_{\mathcal{S}} \leftarrow n_{\mathcal{S}} - 1$
        - $n_{\mathcal{E}} \leftarrow n_{\mathcal{E}} + 1$
        - break
      - **if** tmp_pop[j] = $\mathcal{E} \wedge \|i, j\| \leq r_{\mathcal{E}} \wedge$ random() $\leq p_{\mathcal{E}}$ **then**
        - pop[i] $\leftarrow \mathcal{E}$ ▷ *$\mathcal{S} \to \mathcal{E}$ after contacting with $\mathcal{E}$*
        - $n_{\mathcal{S}} \leftarrow n_{\mathcal{S}} - 1$
        - $n_{\mathcal{E}} \leftarrow n_{\mathcal{E}} + 1$
        - break
  - **if** tmp_pop[i] = $\mathcal{E} \wedge$ random() $< 1/\tau_{\mathcal{E}}$ **then**
    - pop[i] $\leftarrow \mathcal{I}$ ▷ *$\mathcal{E} \to \mathcal{I}$*
    - $n_{\mathcal{E}} \leftarrow n_{\mathcal{E}} - 1$
    - $n_{\mathcal{I}} \leftarrow n_{\mathcal{I}} + 1$
  - **if** tmp_pop[i] = $\mathcal{I} \wedge$ random() $< 1/\tau_{\mathcal{I}}$ **then**
    - pop[i] $\leftarrow \mathcal{R}$ ▷ *Recovered*
    - $n_{\mathcal{I}} \leftarrow n_{\mathcal{I}} - 1$
    - $n_{\mathcal{R}} \leftarrow n_{\mathcal{R}} + 1$
  - **if** tmp_pop[i] = $\mathcal{S} \vee$ tmp_pop[i] = $\mathcal{R}$ **then**
    - **if** random() $< f_G$(age[i]) **then**
      - pop[i] $\leftarrow \mathcal{S}$ ▷ *Removed...*
      - age[i] $\leftarrow$ 0 ▷ *...then Susceptible*
      - **if** tmp_pop[i]=$\mathcal{S}$ **then** $n_{\mathcal{S}} \leftarrow n_{\mathcal{S}} - 1$
      - **if** tmp_pop[i]=$\mathcal{R}$ **then** $n_{\mathcal{R}} \leftarrow n_{\mathcal{R}} - 1$
      - $n_{\mathcal{S}} \leftarrow n_{\mathcal{S}} + 1$
  - **else** ▷ *tmp_pop[i] = $\mathcal{E} \vee$ tmp_pop[i] = $\mathcal{I}$*
    - **if** random() $< f_C$(age[i]) **then**
      - pop[i] $\leftarrow \mathcal{S}$ ▷ *Removed...*
      - age[i] $\leftarrow$ 0 ▷ *...then Susceptible*
      - **if** tmp_pop[i]=$\mathcal{E}$ **then** $n_{\mathcal{E}} \leftarrow n_{\mathcal{E}} - 1$
      - **if** tmp_pop[i]=$\mathcal{I}$ **then** $n_{\mathcal{I}} \leftarrow n_{\mathcal{I}} - 1$
      - $n_{\mathcal{S}} \leftarrow n_{\mathcal{S}} + 1$
      - $n_{\mathcal{D}} \leftarrow n_{\mathcal{D}} + 1$ ▷ *cumulative number of deaths caused by infections*
  - return$(t, n_{\mathcal{S}}/L^2, n_{\mathcal{E}}/L^2, n_{\mathcal{I}}/L^2, n_{\mathcal{R}}/L^2, n_{\mathcal{D}}/L^2)$

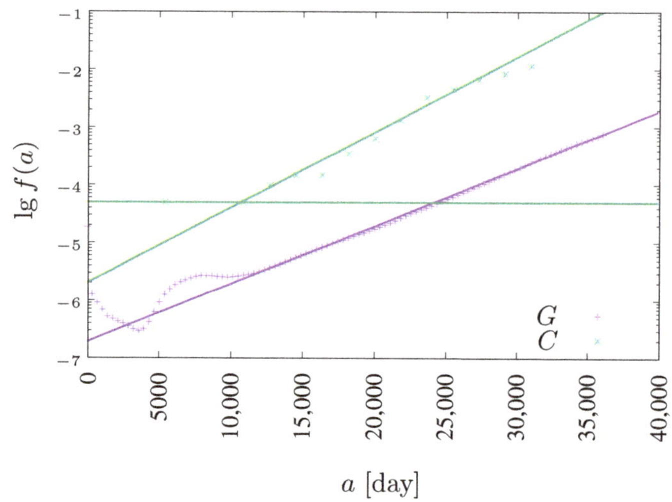

Figure 2. Daily death probability $f(a)$ for patients infected by the coronavirus (SARS-CoV-2, $\times$, $f_C(a)$, [50]) and natural death probability ($+$, $f_G(a)$, [44]).

3. Results

In Figure 3 snapshots from a single-run simulation are presented. They give a quantitative picture of the influence of the interaction range (neighborhood radius) on the spread of the disease. The snapshots in Figure 3a–e show the situation for fixed parameters $p_{\mathcal{E}} = 0.03$ and $p_{\mathcal{I}} = 0.02$ at the $t = 150$ time step, which corresponds to five months after introducing (at random site) 'Patient Zero'. The last subfigure (Figure 3f) presents a situation after a very long time of simulations ($t > 20{,}000$) where the recovered agents die due to their age (according to Equation (5)) and are subsequently replaced by newly born children. For the interaction limited to the first coordination zone (Figure 1a) the disease propagation stays limited to the nearest neighbors of 'Patient Zero'. On the other hand, for the neighborhood with sites up to the fifth coordination zone (Figure 1e) for the same infection rates ($p_{\mathcal{E}} = 0.03$ and $p_{\mathcal{I}} = 0.02$) the disease affects all agents in the population (see Figure 3e). The direct evolution of the system based on [51] can be simulated and observed with the JavaScript application available at [52].

In Figure 4 the fraction $n_{\mathcal{I}}$ of infected agents (in state $\mathcal{I}$) is presented. The figure shows the results of ten different simulations for values of the neighborhood radius $r_{\mathcal{E}} = r_{\mathcal{I}} = 1.5$, $p_{\mathcal{E}} = 0.03$, $p_{\mathcal{I}} = 0.02$. In addition, the results of averaging over $R = 10$ simulations are presented. In two out of ten cases, the epidemic died out right after the start, while in the remaining eight cases it lasted from about eight hundred to over a thousand time steps (days). The figure also shows that the averaging of the results allows for a significant smoothing of the curves, which fluctuate strongly for individual simulations. Based on this test (for (not shown) roughly doubly large statistics, $R = 25$, which do not reveal significant deviations), we decided to average our results (presented in Figures 5–7) over ten independent simulations.

The diagrams in Figures 5 and 6 show the evolution of the epidemic. Namely, they show the number of agents in each state on each day of the epidemic, as well as the cumulative fraction $n_{\mathcal{D}}$ of deaths ($\mathcal{D}$). The fraction $n_{\mathcal{S}}$ of susceptible agents and the fraction $n_{\mathcal{R}}$ of recovered agents are shown on the left vertical axis, while the fractions $n_{\mathcal{E}}$ of exposed agents and $n_{\mathcal{I}}$ of infected agents and the cumulative fraction $n_{\mathcal{D}}$ of deaths caused by infection are shown on the right vertical axis.

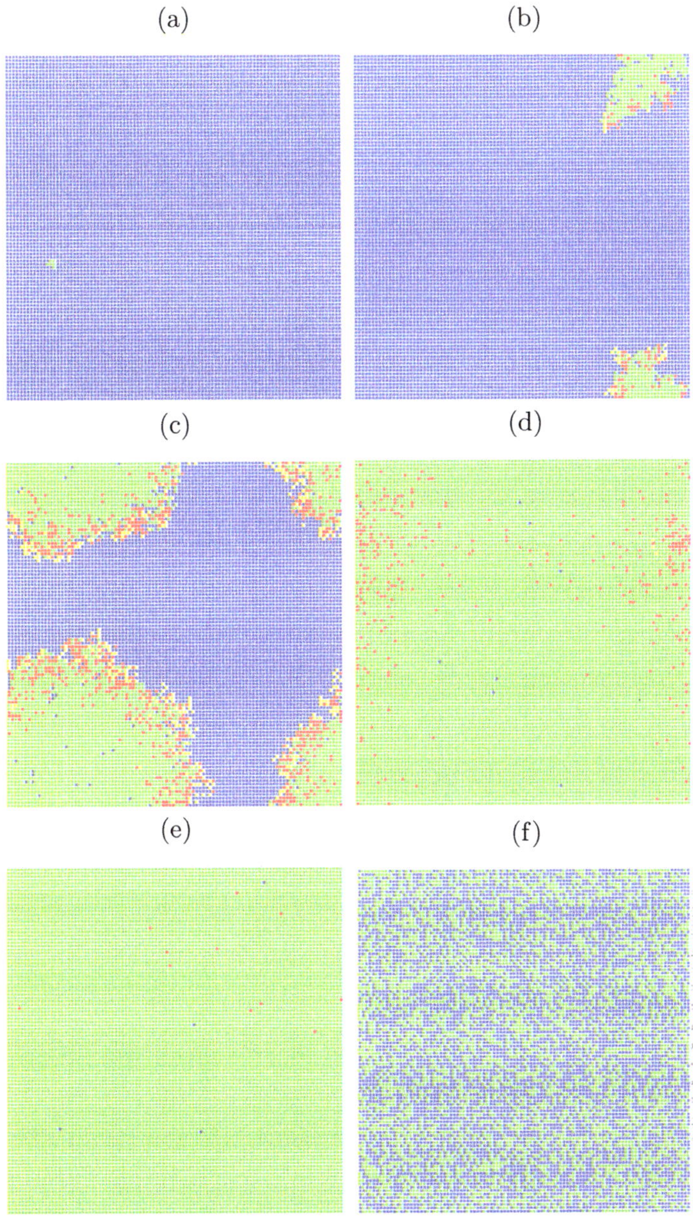

Figure 3. Snapshots from direct simulation [52] for $p_{\mathcal{E}} = 0.03$, $p_{\mathcal{I}} = 0.02$, $R = 1$. The assumed ranges of interactions are (**a**) $r_{\mathcal{E}} = r_{\mathcal{I}} = 1$, (**b**) $r_{\mathcal{E}} = r_{\mathcal{I}} = 1.5$, (**c**) $r_{\mathcal{E}} = r_{\mathcal{I}} = 2$, (**d**,**f**) $r_{\mathcal{E}} = r_{\mathcal{I}} = 2.5$, and (**e**) $r_{\mathcal{E}} = r_{\mathcal{I}} = 3$. The simulation took $t = 150$ time steps except for Figure 3f, where the situation after $t > 20{,}000$ time steps is presented.

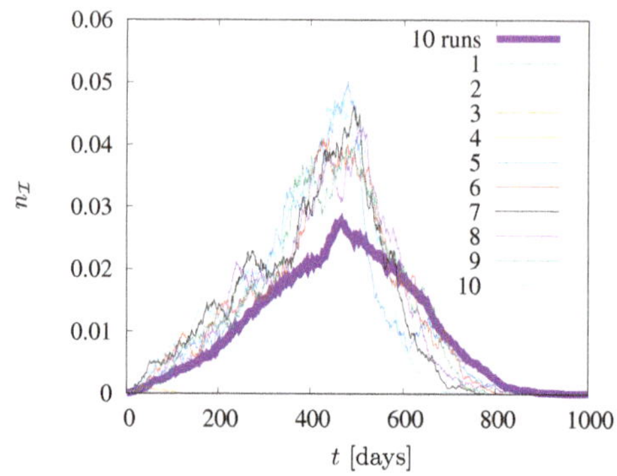

Figure 4. Ten different simulations for values of the neighborhood radius $r_\mathcal{E} = r_\mathcal{I} = 1.5$. $p_\mathcal{E} = 0.03$, $p_\mathcal{I} = 0.02$.

3.1. $r_\mathcal{E} = r_\mathcal{I} = 0$

The case of $r_E = r_I = 0$ (corresponding to total lockdown) leads to immediate disease dieout as only n_0L^2 'Patients Zero' at $t = 0$ are infected and recover after about $\sim 1/q_C(a)$ time steps (days) depending on the agent's age a.

3.2. $r_\mathcal{E} = r_\mathcal{I} > 0$

In Figure 5 the dynamics of states fractions n for various values of the neighborhood radius $r_\mathcal{E} = r_\mathcal{I}$ are presented. We assume infection rates $p_\mathcal{E} = p_\mathcal{I} = 0.005$. The assumed transition rates $p_\mathcal{E}$ and $p_\mathcal{I}$ are very low. As a result, the disease has a very limited chance of spreading in society.

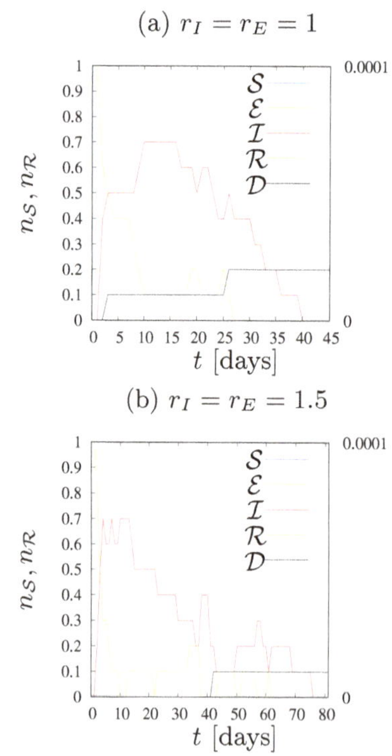

Figure 5. *Cont.*

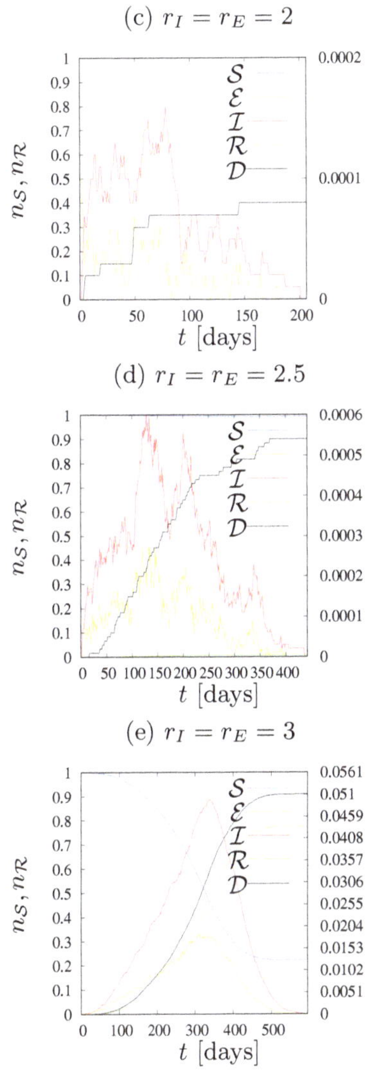

Figure 5. Dynamics of states fractions for various values of the neighborhood radius $r_\mathcal{E} = r_\mathcal{I}$. $p_\mathcal{E} = p_\mathcal{I} = 0.005$, $R = 10$.

Figure 5a illustrates the situation where an infected person (independently either in the $\mathcal{E}$ or $\mathcal{I}$ states) can only infect the four closest neighbors. The epidemic lasted a maximum of forty days, the number of agents who were ill at the same time was less than one (on average, in ten simulations), there were only two deaths out of ten simulations, and the population was not affected by the disease.

The case presented in Figure 5b illustrates the situation in which each person can infect up to eight neighbors. This does not cause significant changes during the epidemic compared with Figure 5a; the average number of simultaneously ill agents remained below one, this time only one person in ten simulations died due to SARS-CoV-2 infection, and the duration of the epidemic was approximately 75 days—twice longer than presented in Figure 5a. We would like to emphasize that the term 'duration of the epidemic' determines the time of the longest duration of the epidemic among the ten simulations carried out.

Figure 5c shows the situation where there were twelve agents in the neighborhood of each cell. The transition rates ($p_{\mathcal{E}}$ and $p_{\mathcal{I}}$) turned out to be so low, that—despite extending the neighborhood—the epidemic vanished quickly. This time the fractions of exposed and infected agents were slightly higher, the maximum number of sick agents in one day was more than one, the longest simulation lasted 80 days, and four agents died within ten iterations.

Increasing the radius of the neighborhood to 2.5 (see Figure 5d) increased the number of exposed and infected agents more than twice compared to Figure 5c. On the day of the peak of the epidemic, five agents were sick and seven died during the epidemic (on average). The epidemic lasted about 600 days, but only about 1–2% of the population became infected throughout the epidemic.

Only an increase in the number of agents in the neighborhood to 24 (as presented in Figure 5e) caused a smooth and rapid development of the epidemic. In this case, it lasted about 550 days, the largest number of sick agents in one day was approximately 500, and the same number of agents also died throughout the epidemic. During the epidemic, around 75% of the population became infected.

In the left column of Figure 6 the dynamics of states fractions n for various values of the radius of the neighborhood $r_{\mathcal{E}} = r_{\mathcal{I}}$ are presented. We assume $p_{\mathcal{E}} = 0.03$, $p_{\mathcal{I}} = 0.02$. As we still keep $r_{\mathcal{I}} = r_{\mathcal{E}}$, setting $p_{\mathcal{E}} > p_{\mathcal{I}}$ simulates the fact that exposed agents (who are not aware of their infection) are more dangerous to those around them than those who know that they are sick, and therefore avoid contact if they show severe symptoms of the disease.

In Figure 6a the results for the smallest possible neighborhood radius $r > 0$ is presented. As in the previous cases with $r = 1$, the epidemic stopped quite quickly. The largest number of agents who were ill ($n_{\mathcal{I}} L^2$) at the same time was two, the longest simulation time was 120 days, the average number of deaths was well below one, and totally only a few agents were infected (the curve representing agents in the state of $\mathcal{R}$ is almost not visible).

Figure 6d shows that contact with eight agents in the neighborhood (for assumed values of infection rates $p_{\mathcal{E}}$ and $p_{\mathcal{I}}$) is enough for the pandemic to affect society to a very large extent and last for a long time. The longest simulation took about 1000 days. During the epidemic, almost half of the population was infected, approximately 350 agents died, and at epidemic peak there were just over 180 sick agents in a single day.

Increasing the range of interaction to the next coordination zone (see Figure 6g) caused a significant reduction in the duration of the epidemic, as well as causing greater havoc among agents. In approximately 350 days, roughly 75% of the population was infected, approximately 500 died, and up to 800 were sick on the day when this number was highest.

Increasing the radius to 2.5, as shown in Figure 6j, further shortened the epidemic time, this time to just two hundred days. Almost all agents were infected, slightly more than 600 agents died, and the maximum number of patients on one day was 2000, which is as much as 20% of the population. Every fifth person was infected by the disease on that day, and if one adds about 800 agents who were in the exposed state $\mathcal{E}$, we get a situation where more than one quarter of the population is under the influence of this disease at the same time.

Further increasing the range of interactions (to the fifth coordination zone and 24 neighbors in it, shown in Figure 6m) gave very similar results to the previous one, except that the virus spread even faster, in less than 150 days. The number of deaths was similar (a little below 600), almost all agents had contact with the disease at some stage of the pandemic, and the maximum number of sick agents in one day was 2800. If one adds over 1100 agents in the $\mathcal{E}$ state on the same day, the effect is that the disease affected nearly 40% of agents on the same day.

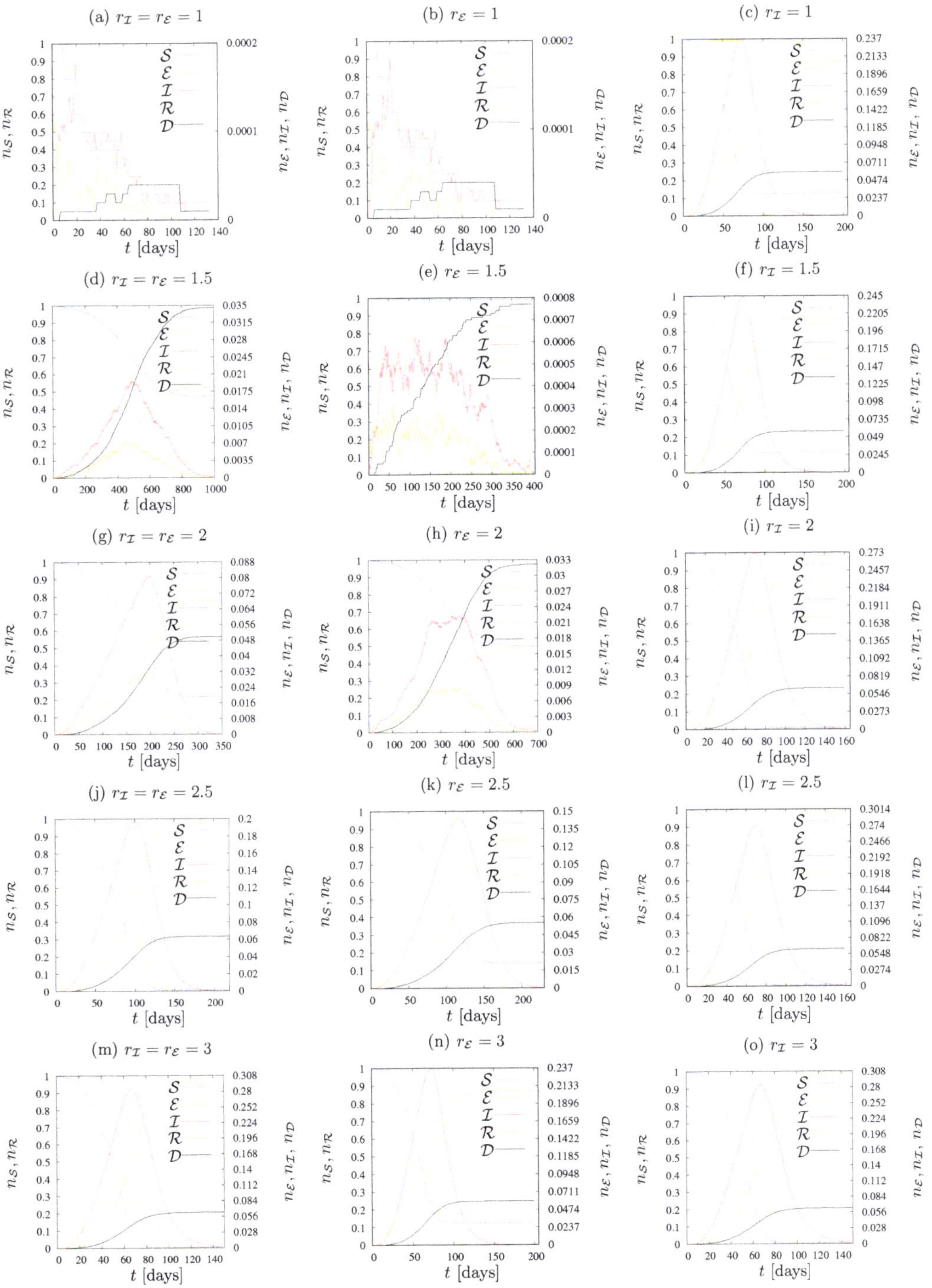

Figure 6. Dynamics of states fractions for various values of the neighborhood radius (**left**) $r_{\mathcal{E}} = r_{\mathcal{I}}$, (**middle**) $r_{\mathcal{E}} \geq r_{\mathcal{I}} = 1$, (**right**) $3 = r_{\mathcal{E}} \geq r_{\mathcal{I}}$. $p_{\mathcal{E}} = 0.03$, $p_{\mathcal{I}} = 0.02$, $R = 10$.

3.3. $r_{\mathcal{E}} \geq r_{\mathcal{I}}$

In the middle column of Figure 6 the dynamics of the state fractions $n_{\mathcal{S}}$, $n_{\mathcal{E}}$, $n_{\mathcal{I}}$, $n_{\mathcal{R}}$ for various values of the neighborhood radius $r_{\mathcal{E}}$ are presented. We assume $p_{\mathcal{E}} = 0.03$ and $p_{\mathcal{I}} = 0.02$. For this set of plots, $r_{\mathcal{I}}$ has been predefined as 1 and only $r_{\mathcal{E}}$ changes. Translating this into a description of the real world, sick agents who have disease symptoms are aware of their illness (infected) and limit their contact with the environment to a minimum, while oblivious agents (exposed) do not realize that they can transmit the disease. In each of the tested parameter sets, agents in the $\mathcal{I}$ state can infect only four of their closest neighbors.

The case illustrated in Figure 6b was analyzed in Section 3.2 in Figure 6a.

The plot in Figure 6e shows the situation where agents in the state $\mathcal{E}$ can infect up to eight agents in their neighborhood. We see irregularities in the shapes of the curves showing the fraction of agents in the states $\mathcal{I}$ and $\mathcal{E}$, and the duration of the epidemic is relatively long, over 400 days. However, the values shown in the graphs show that the pandemic was not dangerous for the entire population. Death from the disease was recorded on average in less than eight agents, the maximum number of sick agents in one day was just over six, and the total number of infected agents in the entire epidemic was so low that the deviation of the curve representing the number of agents in the state $\mathcal{S}$ from the top of the graph is almost imperceptible.

In Figure 6h, the plots for further extension (to the third coordination zone, with $z = 12$ neighbors) of the radius of the neighborhood of exposed agents (in the state $\mathcal{E}$) are presented. The curves are much smoother than those presented in Figure 6e; however, we can still see some irregularities in the curves describing agents in the $\mathcal{E}$ and $\mathcal{I}$ states. The duration of the longest simulation was approximately 700 days, the number of deaths was less than 330, and at the peak of the pandemic, approximately 220 agents were simultaneously ill. The curves presenting the fraction of agents in the states of $\mathcal{S}$ and $\mathcal{R}$ almost perfectly line up on the right side of the chart, meaning that half of the population contracted the disease while the other half remained healthy throughout the epidemic.

The case where agents in the state $\mathcal{E}$ can infect agents within a radius of 2.5 is shown in Figure 6k. The epidemic was definitely more dynamic than in the previously studied case (Figure 6h), lasting only a little over two hundred days, and the maximum number of sick agents in one day $n_{\mathcal{I}}L^2$ reached almost 1500 agents. The sum of those who died from the coronavirus was less than 600, and during the pandemic, approximately 85% of the population became recovered (and earlier exposed and/or infected).

Finally, for the radius $r_{\mathcal{E}}$ increased to three (see Figure 6n) we did not observe too many changes compared to the previous case (Figure 6k). The epidemic was even shorter, it lasted a little more than 150 days, the number of deaths was very similar (it could be estimated at 600), slightly more than 85% of agents in the population were infected, and at the hardest moment of the pandemic there were at the same time about 2370 sick agents (in the $\mathcal{I}$ state). The results are therefore almost identical to those of $r_{\mathcal{E}} = 2.5$, except that disease propagation is faster.

The right column of Figure 6 shows cases with predetermined $r_{\mathcal{E}} = 3$ and various $r_{\mathcal{I}}$. We keep $p_{\mathcal{E}} = 0.03$ and $p_{\mathcal{I}} = 0.02$

All graphs are very similar to each other, as a large range of infections of agents in the state $\mathcal{E}$ makes the influence of $r_{\mathcal{I}}$ on epidemic evolution only marginal. This can be easily observed by comparing the scales presented in all subfigures. The assumed range of infection $r_{\mathcal{E}} = 3$ is large enough (at least for assumed values of transmission rates $p_{\mathcal{E}}$ and $p_{\mathcal{I}}$) to prevent any observable influence of $r_{\mathcal{I}}$ on epidemic trajectories.

4. Discussion

Let us start the discussion with comparing the leftmost (Figure 6a,d,g,j,m) and the middle (Figure 6b,e,h,k,n) columns of Figure 6. The comparison reveals that quarantining or limiting the connectivity of agents in the $\mathcal{I}$ state (both infected and informed) may bring good or even very good results in preventing disease propagation, depending on the arrangement of the other parameters. For $r_{\mathcal{E}} = 1.5$, this completely brought the epidemic to a halt, which would otherwise affect more than half the population. With $r_{\mathcal{E}} = 2$, it makes it possible to reduce the share of infected agents in the population from 75% to 50%. The effects on $r_{\mathcal{E}} = 2.5$ and $r_{\mathcal{E}} = 3$ are similar—instead of the total population, less than 90% of population became infected. When the $r_{\mathcal{E}} = 1.5$ pandemic duration was significantly shortened, it is because the virus was unable to survive, and the disease was extinct. In other cases, the duration of the epidemic increased; the restrictions introduced for agents in the $\mathcal{I}$ state did not completely extinguish the disease, but allowed to slow it down and mitigate its effects.

For low values of the probability of infection ($p_{\mathcal{E}} = p_{\mathcal{I}} = 0.005$), that is, in a situation where the transmission of the virus is not too high, even a slight limitation of the contact among agents allows for a complete inhibition of the disease and protection of the society against its negative effects (see Figure 5). We note that manipulating the $p_{\mathcal{E}}$ and/or $p_{\mathcal{I}}$ parameters may reflect changes in disease transition rates with their low values corresponding to wild variant of the SARS-CoV-2 virus while higher values correspond to the fiercer (including delta and specially omicron) variants of the SARS-CoV-2 virus. For instance, for a fixed radius of interaction ($r_{\mathcal{I}} = r_{\mathcal{E}} = 3$) for low values of $p_{\mathcal{E}} = p_{\mathcal{I}} = 0.005$ nearly 75% of the population reached the $\mathcal{R}$ state (Figure 5e) while increasing infection rates to $p_{\mathcal{E}} = 0.03$ and $p_{\mathcal{I}} = 0.02$ caused the entire population to fall ill (Figure 6m). The summed values of the fractions $n_{\mathcal{E}}$ of exposed and $n_{\mathcal{I}}$ infected agents at the peaks of disease are 6.6% (for $p_{\mathcal{E}} = p_{\mathcal{I}} = 0.005$, Figure 5e) and 39% (for $p_{\mathcal{E}} = 0.03$ and $p_{\mathcal{I}} = 0.02$, Figure 6m) of the population.

The results presented in Figures 5 and 6 are summarized in Figure 7, where the maximum fraction of agents in state $\mathcal{I}$ vs. the neighbours number z for various sets of parameters is presented. We observe a gradual increase in the maximum number of infected agents (up to 30%) as the number of agents in the neighborhood increases. An exception to this rule is observed only when $r_{\mathcal{E}} = 3$, when the maximum level of infections is constant and it does not change with the increase of the range of the interaction of agents in state $\mathcal{I}$.

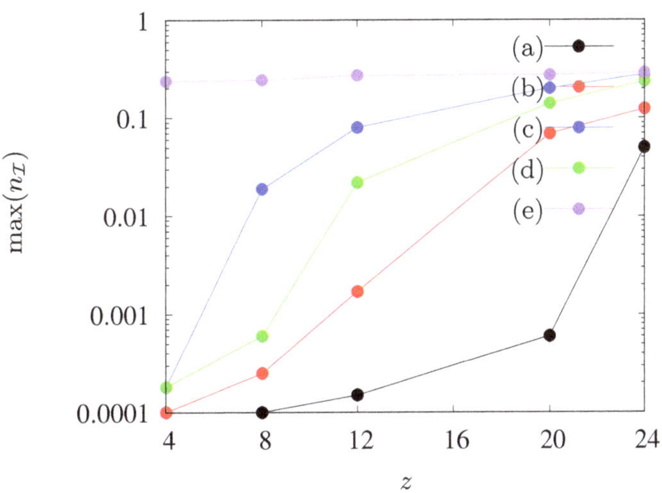

Figure 7. Maximal fraction $n_{\mathcal{I}}$ of agents in state $\mathcal{I}$ as dependent on the number of agents' neighbours z in the neighborhood. (**a**) $p_{\mathcal{E}} = p_{\mathcal{I}} = 0.005$, $z_E = z_{\mathcal{I}} = z$, (**b**) $p_{\mathcal{E}} = p_{\mathcal{I}} = 0.01$, $z_E = z_{\mathcal{I}} = z$, (**c**) $p_{\mathcal{E}} = 0.03$, $p_{\mathcal{I}} = 0.02$, $z_E = z_{\mathcal{I}} = z$, (**d**) $p_{\mathcal{E}} = 0.03$, $p_{\mathcal{I}} = 0.02$, $z_E = z$, $z_{\mathcal{I}} = 4$, (**e**) $p_{\mathcal{E}} = 0.03$, $p_{\mathcal{I}} = 0.02$, $z_E = 24$, $z_{\mathcal{I}} = z$.

5. Conclusions

In this article we present a stochastic synchronous cellular automaton defined on a square lattice. The automaton rules are based on the SEIR model with probabilistic parameters collected from real human mortality data and SARS-CoV-2 disease characteristics.

Automaton rules are presented in Algorithm 1. With computer simulations, we show the influence of the radius of the neighborhood on the number of infected and deceased agents in the artificial population.

The study presented in this paper is based on static automaton. Thus, our approach is equivalent to disease propagation described in terms of conduction-like processes (i.e., the position of each cell is fixed and can infect a neighbor, at distance $r_{\mathcal{E}}$ or $r_{\mathcal{I}}$). The latter is a natural way to model with the cellular automata technique. However, we note that also convection-like processes (i.e., the population can flow within the system) may play a crucial role at both large [53] and local scales [19,54].

Further enrichment of the model can lead to the introduction of additional components, including the compartment $\mathcal{V}$ that describes the vaccinated agents. This model improvement allows us to study scenarios with limited and unlimited vaccine supply [20] or the existence and stability of steady states [55,56]. Vaccinations seem to be particularly effective when the vaccination campaign starts early and with a large number of vaccinated individuals [20]. Moreover, the realistic models should take into account people's attitudes to vaccination programs [57,58].

In conclusion, increasing the radius of the neighborhood (and the number of agents interacting locally) favors the spread of the epidemic. However, for a wide range of interactions of exposed agents, even isolation of infected agents cannot prevent successful disease propagation. This supports aggressive testing against disease as one of the useful strategies to prevent large peaks of infection in the spread of SARS-CoV-2-like disease. The latter can have devastating consequences for the health care system, in particular for the availability of hospital beds for SARS-CoV-2 and other diseases.

Author Contributions: Conceptualization, K.M.; Investigation, S.B.; Methodology, K.M.; Software, S.B.; Visualization, S.B.; Writing-original draft, K.M.; Writing-review & editing, S.B. and K.M. All authors have read and agreed to the published version of the manuscript.

Funding: This research received no external funding.

Data Availability Statement: The theoretical data generated by cellular automata simulations is available from the authors upon a reasonable request. The real-world data is based on [44,50].

Acknowledgments: The authors are grateful to Zdzisław Burda for critical reading of the manuscript and fruitful discussion.

Conflicts of Interest: The authors declare no conflict of interest.

References

1. Zhu, N.; Zhang, D.; Wang, W.; Li, X.; Yang, B.; Song, J.; Zhao, X.; Huang, B.; Shi, W.; Lu, R.; et al. A novel coronavirus from patients with pneumonia in China, 2019. *N. Engl. J. Med.* **2020**, *382*, 727–733. [CrossRef] [PubMed]
2. *WHO COVID-19 Dashboard*; World Health Organization: Geneva, Switzerland, 2020. Available online: https://covid19.who.int/ (accessed on 1 June 2022).
3. Worldometers: COVID-19 Coronavirus Pandemic. 2021. Available online: https://www.worldometers.info/coronavirus/ (accessed on 1 December 2021).
4. Zhang, Y.; Gong, C.; Li, D.; Wang, Z.-W.; Pu, S.D.; Robertson, A.W.; Yu, H.; Parrington, J. A prognostic dynamic model applicable to infectious diseases providing easily visualized guides: A case study of COVID-19 in the UK. *Sci. Rep.* **2021**, *11*, 8412. [CrossRef] [PubMed]
5. Lima, L.L.; Atman, A.P.F. Impact of mobility restriction in COVID-19 superspreading events using agent-based model. *PLoS ONE* **2021**, *16*, e0248708. [CrossRef] [PubMed]
6. Medrek, M.; Pastuszak, Z. Numerical simulation of the novel coronavirus spreading? *Expert Syst. Appl.* **2021**, *166*, 114109. [CrossRef]
7. Schimit, P.H.T. A model based on cellular automata to estimate the social isolation impact on COVID-19 spreading in Brazil. *Comput. Methods Programs Biomed.* **2021**, *200*, 105832. [CrossRef]
8. Dai, J.; Zhai, C.; Ai, J.; Ma, J.; Wang, J.; Sun, W. Modeling the spread of epidemics based on cellular automata. *Processes* **2021**, *9*, 55. [CrossRef]
9. Zupanc, G.K.H.; Lehotzky, D.; Tripp, I.P. The neurosphere simulator: An educational online tool for modeling neural stem cell behavior and tissue growth. *Dev. Biol.* **2021**, *469*, 80–85. [CrossRef]
10. Gwizdalla, T. Viral disease spreading in grouped population. *Comput. Methods Progrems Biomed.* **2020**, *197*, 105715. [CrossRef]

11. Monteiro, L.H.A.; Fanti, V.C.; Tessaro, A.S. On the spread of SARS-CoV-2 under quarantine: A study based on probabilistic cellular automaton. *Ecol. Complex.* **2020**, *44*, 100879. [CrossRef]
12. Ghosh, S.; Bhattacharya, S. A data-driven understanding of COVID-19 dynamics using sequential genetic algorithm based probabilistic cellular automata. *Appl. Soft Comput.* **2020**, *96*, 106692. [CrossRef]
13. Zhou, Y.; Li, J.; Chen, Z.; Luo, Q.; Wu, X.; Ye, L.; Ni, H.; Fei, C. The global COVID-19 pandemic at a crossroads: Relevant countermeasures and ways ahead. *J. Thorac. Dis.* **2020**, *12*, 5739–5755. [CrossRef] [PubMed]
14. Mondal, S.; Mukherjee, S.; Bagchi, B. Mathematical modeling and cellular automata simulation of infectious disease dynamics: Applications to the understanding of herd immunity. *J. Chem. Phys.* **2020**, *153*, 114119. [CrossRef] [PubMed]
15. Tang, L.; Zhou, Y.; Wang, L.; Purkayastha, S.; Zhang, L.; He, J.; Wang, F.; Song, P.X.-K. A review of multi-compartment infectious disease models. *Int. Stat. Rev.* **2020**, *88*, 462–513. [CrossRef] [PubMed]
16. Yuan, M. Geographical information science for the United Nations' 2030 agenda for sustainable development. *Int. J. Geogr. Inf. Sci.* **2021**, *35*, 1–8. [CrossRef]
17. Dascalu, M.; Malita, M.; Barbilian, A.; Franti, E.; Stefan, G.M. Enhanced cellular automata with autonomous agents for COVID-19 pandemic modeling. *Rom. J. Inf. Sci. Technol.* **2020**, *23*, S15–S27.
18. Orzechowska, J.; Fordon, D.; Gwizdałła, T. Size effect in cellular automata based disease spreading model. *Lect. Notes Comput. Sci.* **2018**, *11115*, 146–153.
19. Nava, A.; Papa, A.; Rossi, M.; Giuliano, D. Analytical and cellular automaton approach to a generalized SEIR model for infection spread in an open crowded space. *Phys. Rev. Res.* **2020**, *2*, 043379. [CrossRef]
20. Gabrick, E.C.; Protachevicz, P.R.; Batista, A.M.; Iarosz, K.C.; de Souza, S.L.; Almeida, A.C.; Szezech, J.D.; Mugnaine, M.; Caldas, I.L. Effect of two vaccine doses in the SEIR epidemic model using a stochastic cellular automaton. *Phys. A Stat. Mech. Its Appl.* **2022**, *597*, 127258. [CrossRef]
21. Ilachinski, A. *Cellular Automata: A Discrete Universe*; World Scientific: Singapore, 2001.
22. Wolfram, S. *A New Kind of Science*; Wolfram Media. Available online: https://www.wolfram-media.com/ (accessed on 1 December 2021).
23. Chopard, B.; Droz, M. *Cellular Automata Modeling of Physical Systems*; Cambridge University Press: Cambridge, UK, 2005.
24. Chopard, B. *Computational Complexity: Theory, Techniques, and Applications*; Meyers, R.A., Ed.; Springer: New York, NY, USA, 2012; pp. 407–433.
25. de Oliveira, P.M.C. Epidemics a la Stauffer. *Phys. A Stat. Mech. Its Appl.* **2021**, *561*, 125287. [CrossRef]
26. Lux, T. The social dynamics of COVID-19. *Phys. A Stat. Mech. Its Appl.* **2021**, *567*, 125710. [CrossRef]
27. Weisbuch, G. Urban exodus and the dynamics of COVID- 19 pandemics. *Phys. A Stat. Mech. Its Appl.* **2021**, *569*, 125780. [CrossRef] [PubMed]
28. Lorig, F.; Johansson, E.; Davidsson, P. Agent-based social simulation of the COVID-19 pandemic: A systematic review. *J. Artif. Soc. Soc. Simul.* **2021**, *24*, 5. [CrossRef]
29. Ziff, R.M. Percolation and the pandemic. *Phys. A Stat. Mech. Its Appl.* **2021**, *568*, 125723. [CrossRef]
30. Ross, R. An application of the theory of probabilities to the study of a priori pathometry—Part I. *Proc. R. Soc. Lond.* **1916**, *92*, 204–230.
31. Ross, R.; Hudson, H.P. An application of the theory of probabilities to the study of a priori pathometry—Part II. *Proc. R. Soc. Lond.* **1917**, *93*, 212–225.
32. Ross, R.; Hudson, H.P. An application of the theory of probabilities to the study of a priori pathometry—Part III. *Proc. R. Soc. Lond.* **1917**, *93*, 225–240.
33. Kermack, W.O.; McKendrick, A.G. A contribution to the mathematical theory of epidemics. *Proc. R. Soc. Lond.* **1927**, *115*, 700–721.
34. Kermack, W.O.; McKendrick, A.G. Contributions to the mathematical theory of epidemics—I. *Bull. Math. Biol.* **1991**, *53*, 33–55.
35. Kendall, D.G. *Berkeley Symposium on Mathematical Statistics and Probability*; University of California Press: Berkeley, CA, USA, 1956; pp. 149–165.
36. Hethcote, H.W. The mathematics of infectious diseases. *SIAM Rev.* **2000**, *42*, 599–653. [CrossRef]
37. Newman, M.E.J. Spread of epidemic disease on networks. *Phys. Rev. E* **2002**, *66*, 016128. [CrossRef]
38. Worldometers: COVID-19 Coronavirus Pandemic. 2021. Available online: https://www.worldometers.info/coronavirus/coronavirus-symptoms/ (accessed on 1 December 2021).
39. Worldometers: COVID-19 Coronavirus Pandemic. 2021. Available online: https://www.worldometers.info/coronavirus/coronavirus-incubation-period/ (accessed on 1 December 2021).
40. Barthélemy, M. Spatial networks. *Phys. Rep.* **2011**, *499*, 1–101. [CrossRef]
41. Burda, Z. Modelling excess mortality in COVID-19-like epidemics. *Entropy* **2020**, *22*, 1236. [CrossRef] [PubMed]
42. Dureau, J.; Kalogeropoulos, K.; Baguelin, M. Capturing the time-varying drivers of an epidemic using stochastic dynamical systems. *Biostatistics* **2013**, *14*, 541–555. [CrossRef]
43. Faranda, D.; Alberti, T. Modeling the second wave of COVID-19 infections in France and Italy via a stochastic SEIR model. *Chaos Interdiscip. J. Nonlinear Sci.* **2020**, *30*, 111101. [CrossRef] [PubMed]
44. Arias, E. United States life tables, 2003. *Natl. Vital Stat. Rep.* **2006**, *54*, 1–40. [PubMed]
45. Richmond, P.; Roehner, B.M.; Irannezhad, A.; Hutzler, S. Mortality: A physics perspective. *Phys. A Stat. Mech. Its Appl.* **2021**, *566*, 125660. [CrossRef]

46. Gompertz, B. On the nature of the function expressive of the law of human mortality, and on a new mode of 11 determining the value of life contingencies. *Philos. Trans. R. Soc. Lond.* **1825**, *115*, 513– 583.
47. Vandamme, L.K.J.; Rocha, P.R.F. Analysis and simulation of epidemic COVID-19 curves with the Verhulst model applied to statistical inhomogeneous age groups. *Appl. Sci.* **2021**, *11*, 4159. [CrossRef]
48. Filho, T.M.R.; Moret, M.A.; Chow, C.C.; Phillips, J.C.; Cordeiro, A.J.A.; Scorza, F.A.; Almeida, A.-C.G.; Mendes, J.F.F. A data-driven model for COVID-19 pandemic—Evolution of the attack rate and prognosis for Brazil. *Chaos Solitons Fractals* **2021**, *152*, 111359. [CrossRef]
49. Makowiec, D.; Stauffer, D.; Zielinski, M. Gompertz law in simple computer model of aging of biological population. *Int. J. Mod. Phys. C* **2001**, *12*, 1067–1073. [CrossRef]
50. Distribution of Deaths Due to the Coronavirus (COVID-19) in Poland as of January 2021, by Age Group. Data Based on National Institute of Public Health PZH Report (in Polish). 2021. Available online: https://www.statista.com/statistics/1110890/poland-coronavirus-covid-19-fatalities-by-age/ (accessed on 1 March 2022). [CrossRef]
51. Biernacki, S. Computer Simulation of the Impact of Quarantine and Limitation of Long-Range Communication on the Spread of an Epidemic of a Drop-Borne Virus. Master's Thesis, AGH University of Science and Technology, Kraków, Poland, 2021.
52. Biernacki, S. Javascript Application. Available online: http://www.zis.agh.edu.pl/app/MSc/Szymon_Biernacki/ (accessed on 1 December 2021).
53. Belik, V.; Geisel, T.; Brockmann, D. Natural human mobility patterns and spatial spread of infectious diseases. *Phys. Rev. X* **2011**, *1*, 011001. [CrossRef]
54. Mello, B.A. One-way pedestrian traffic is a means of reducing personal encounters in epidemics. *Front. Phys.* **2020**, *8*, 376. [CrossRef]
55. Sun, D.; Li, Y.; Teng, Z.; Zhang, T.; Lu, J. Dynamical properties in an SVEIR epidemic model with age-dependent vaccination, latency, infection, and relapse. *Math. Methods Appl. Sci.* **2021**, *44*, 12810–12834. [CrossRef]
56. Nabti, A.; Ghanbari, B. Global stability analysis of a fractional SVEIR epidemic model. *Math. Methods Appl. Sci.* **2021**, *44*, 8577–8597. [CrossRef]
57. Bier, M.; Brak, B. A simple model to quantitatively account for periodic outbreaks of the measles in the Dutch Bible Belt. *Eur. Phys. J.* **2015**, *88*, 107. [CrossRef]
58. Lisowski, B.; Yuvan, S.; Bier, M. Outbreaks of the measles in the Dutch Bible Belt and in other places—New prospects for a 1000 year old virus. *Biosystems* **2019**, *177*, 16–23. [CrossRef]

entropy

MDPI

Article

Proof of Equivalence of Carnot Principle to II Law of Thermodynamics and Non-Equivalence to Clausius I and Kelvin Principles

Grzegorz Marcin Koczan

Department of Mechanical Processing of Wood, Institute of Wood Sciences and Furniture, Warsaw University of Life Sciences, 02-776 Warsaw, Poland; grzegorz_koczan@sggw.edu.pl

Abstract: The II law of thermodynamics is most often given in three supposedly equivalent formulations: two Clausius (I and II) and one Kelvin. The most general and indisputable entropy formulation belongs to Clausius (II). The earlier Clausius I principle determines the natural direction heat flow between bodies at different temperatures. On the other hand, the Kelvin principle states that it is impossible to completely convert heat into work. The author argues that the Kelvin principle is a weaker statement (or more strictly non-equivalent) than the Clausius I principle, and the latter is a weaker statement than Carnot principle, which is equivalent to Clausius II principle. As a result, the Kelvin principle and the Clausius I principle are not exhaustive formulations of the II law of thermodynamics. At the same time, it turns out that the Carnot principle becomes such a formulation. Apart from providing a complete set of proof and disproof, the author, indicates where the methodological errors were made in the alleged proof of the equivalence of the Kelvin principle and both Clausius principles.

Keywords: II law of thermodynamics; Carnot principle; Kelvin principle; Ostwald principle; *perpetuum mobile* type III; Clausius I and II principles; formal implication; model theory

Citation: Koczan, G.M. Proof of Equivalence of Carnot Principle to II Law of Thermodynamics and Non-Equivalence to Clausius I and Kelvin Principles. *Entropy* **2022**, *24*, 392. https://doi.org/10.3390/e24030392

Academic Editor: Adam Gadomski

Received: 13 February 2022
Accepted: 7 March 2022
Published: 11 March 2022

1. Introduction

We shall start with the chronological formulation of the fundamental principles of thermodynamics according to Carnot, Clausius I, Kelvin and Clausius II. These four principles are closely related to the II law of thermodynamics. It can be said that the Carnot principle opened the way to the formulation of the II law of thermodynamics, and the Clausius II principle fully achieved this goal. It is commonly claimed (allegedly proven) that the principles of Clausius I and II and the Kelvin principles are equivalent, and thus they constitute the II law of thermodynamics. On the other hand, the Carnot principle is claimed to be only a consequence of the II law of thermodynamics, not its equivalent formulation. Therefore, the relationship between all four principles needs to be verified.

In order for the formulated principles to be exact, it is necessary to define the context: the virtual space of thermodynamic processes and the elementary rules by which they are governed. The processes will be represented on the basis of diagrams characteristic of heat engines or heat pumps (see Figure 1). Such diagrams include a heat reservoir (thermostat) with a temperature of T_1 (heater) and a heat reservoir (thermostat) with a temperature of T_2 (cooler, $T_2 < T_1$). The reservoirs can exchange heat naturally between each other or with a device that can perform work (engine) or, at the expense of work, force unnatural heat transfer between the reservoirs (heat pump or refrigerator). The diagrams can also show impossible virtual thermodynamic processes. The role of the formulated thermodynamic principles is to narrow the set of thermodynamic processes to a set of physically possible processes. Nevertheless, the initial Ω set of all considered diagrams (processes) must also include non-physical processes. Otherwise, it would not be possible to prove or disprove the implications of the two principles, which are often based on the negation of physical processes. The set of diagrams of Ω virtual processes will, however, be limited by the assumption that the I law of thermodynamics (or more precisely, the law of energy conservation) is met ($Q_1 = W + Q_2$).

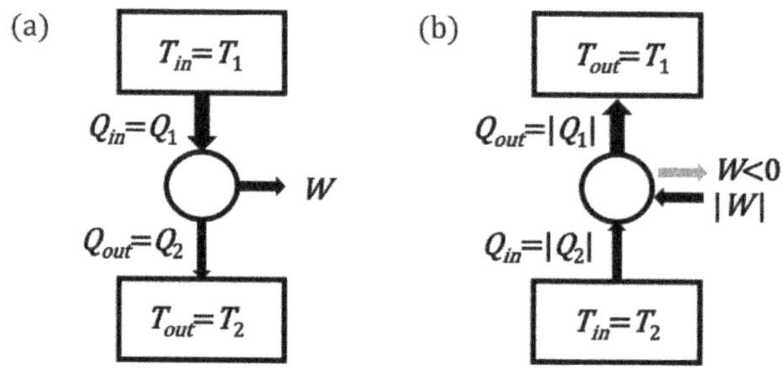

Figure 1. Schemes of two basic thermal devices: (**a**) A typical heat engine operating at the expense of the heat from the higher temperature reservoir. (**b**) A device that extracts heat from the lower temperature reservoir is necessarily (e.g., in the Carnot principle) a heat pump or a refrigerator that must be powered by work.

The simplicity of the diagrams allows them to be added naturally, reflecting the superposition of physical processes. Such a superposition appears implicitly in any textbook proof of the equivalence of the thermodynamic principles which are under consideration in this work. It is, therefore, natural that where two diagrams comply with a given principle, their sum should also comply with that principle. In other words, it is assumed that the set of diagrams in accordance with a given physical principle should be a convex set in the above-mentioned sense.

Carnot principle (*C*0)
The efficiency of the heat conversion process $Q_{in} > 0$ to work W in the device operating in the range between the temperature of the heat source T_{in} and the temperature of the heat receiver T_{out}, cannot be greater than the ratio of the difference of these temperatures to the temperature of the heat source:

$$\frac{W}{Q_{in}} \leq \frac{T_{in} - T_{out}}{T_{in}}. \tag{1}$$

Such a formulation does not explicitly contain the information that the temperature of the heat source must be higher than that of the heat receiver. However, the implicit condition that the engine performs real positive work ($W > 0$) entails the aforementioned temperatures relationship $T_{in} = T_1 > T_{out} = T_2$. In this case, the condition of the Carnot principle takes the form:

$$\eta := \frac{W}{Q_1} \leq \frac{T_1 - T_2}{T_1} =: \eta_C. \tag{2}$$

Mathematically speaking, we can also consider the process of work performed by an external force ($W < 0$), then $T_{in} = T_2 < T_{out} = T_1$ and:

$$\frac{W}{\mid Q_2 \mid} \leq \frac{T_2 - T_1}{T_2} < 0. \tag{3}$$

This correct condition applies to refrigeration processes in which, thanks to external work, it is possible to pump heat from a lower to a higher temperature. Such processes exhibit high coefficients of cooling efficiency $\mid Q_2 \mid / \mid W \mid$ or the efficiency of the heat pump $\mid Q_1 \mid / \mid W \mid \geq 1$. However, it is convenient to introduce an efficiency parameter $\tilde{\eta} \leq 1$ for the refrigeration processes, which corresponds to an engine process running in the opposite direction. Thus, such a parameter is determined analogously to (2), with the sense that the minuses are shortened $\tilde{\eta} = \mid W \mid / \mid Q_1 \mid = (-W)/(-Q_1) = W/Q_1$. To find the condition for $\tilde{\eta}$, let us transform (3) by virtue of $W = \mid Q_2 \mid - \mid Q_1 \mid < 0$:

$$1 - \frac{\mid Q_1 \mid}{\mid Q_2 \mid} \leq 1 - \frac{T_1}{T_2} \Rightarrow \frac{\mid Q_2 \mid}{\mid Q_1 \mid} \leq \frac{T_2}{T_1} \Rightarrow \tilde{\eta} \geq \eta_C. \quad (4)$$

Thus, the efficiency parameter of $\tilde{\eta}$ cooling processes is not lower than the Carnot efficiency (unlike the engine processes which have the efficiency not greater than the Carnot efficiency). This result was obtained here on the basis of the application of the Carnot principle for negative work $W < 0$. This result will be later deduced through independent considerations of a different kind.

Historically speaking, Carnot in 1824 [1] did not yet know the concept of absolute temperature and only defined the dependence of efficiency on a temperature difference [2]. Nevertheless, his considerations on the isothermal-adiabatic cycle implied the existence of maximum efficiency and minimal temperature. This minimum temperature along with the absolute temperature scale were formally introduced by Thomson (Lord Kelvin) [3]. Such absolute temperature scale is adopted in this work. This temperature is consistent with the zeroth law of thermodynamics, and its changes are consistent with the operation of a thermometer—so there is no need to give its elementary definition. The expression for the efficiency $\eta_C(T_1, T_2)$ of the Carnot cycle was known as the Carnot function, which was not originally fully defined by Carnot. This function was intensely sought after by Clausius and Kelvin. Thus, the above-mentioned scholars, who had always referred to Carnot, also made an important contribution to the contemporary form of the C0 principle [4].

Carnot, as a pioneer of thermodynamics [3,5,6], did not know yet either the law of conservation of energy or the I law of thermodynamics [7], so he used the concept of the flow of the indestructible fluid of heat (caloric) [2]. Nevertheless, the Carnot principle can be considered in the context of the II law of thermodynamics [8–10], putting aside the fact that Carnot did not know the I law of thermodynamics. In the context of the [11] research of the finite-time Carnot cycle, it can be concluded that the idea of a Carnot engine, in a sense, is still not a closed concept in terms of engineering and experience.

Clausius I principle (CI)
Heat naturally flows from a body at a higher temperature to a body at a lower temperature. Therefore, a direct (not forced by work) process of heat transfer from the body at a lower temperature to the body at a higher temperature is not possible.

An absolute temperature scale is not needed to determine which reservoir has a higher temperature—even the Celsius scale is sufficient here. The Clausius I principle seems so trivial that one might suppose that it has no far-reaching, less trivial consequences. Nevertheless, it is allegedly proved, that the Clausius I principle is equivalent to the full II law of thermodynamics [12–15]. Criticism of this popular proof could only be found in a 2008 preprint by a scholar Bhattacharyya [16].

The CI principle is ascribed to Clausius' publications from 1850, 1851 [17,18] and 1854, 1856 [19,20]. Contemporary work by Xue and Guo suggests that the above assignment is not strict [21]. These authors do not distinguish two Clausius principles (CI, CII), so it is not surprising that their reference to later Clausius textbooks [22,23] has to do with the second version of the Clausius principle (CII). However, in the chapter: *"III. Second Main Principle of the Mechanical Theory of Heat"* and in section: *"5. New Fundamental Principle concerning Heat"* of textbook [23], Clausius wrote: *"...and then enunciate the principle as follows: 'A passage of heat from a colder to a hotter body cannot take place without compensation'. This proposition, laid down as a Fundamental Principle by the author..."* Such a declaration by Clausius from 1879 constitutes his principle in the first version (CI).

Kelvin principle (K).
The processes of converting heat to work and work to heat do not run symmetrically. A full conversion of work to heat (internal energy) is possible. However, a full conversion of the heat to the work is not possible in a cyclical process.

The principle called the Kelvin principle is formulated in various but similar ways. This relates not so much to the original formulation of Kelvin [24,25], but more to a later formulation of Planck (see [26]), called the Kelvin–Planck formulation (see [27]). The above version of the K principle reflects, as intended by the author, the essence of all these formulations—although this form of the principle comes from Planck or Ostwald ([26]) rather than initially from Kelvin. Nevertheless, traditional formulations identically equivalent to the K version of the principle are called the Kelvin principle, not the Kelvin–Planck or Ostwald principle [12–15,28–31]. In the publications [24,25,32], Kelvin expressed his principle as follows: *"It is impossible, by means of inanimate material agency, to derive mechanical effect from any portion of matter by cooling it below the temperature of the coldest of the surrounding objects"*. Such a formulation, however, seems weaker than the formulation of K, as it seems to refer to the performance of work at the expense of the cooler heat, and not more naturally at the expense of the heater heat. Therefore, it is not known which of the following elements Kelvin wanted to emphasize in this formulation: (i) a change in the temperature of the reservoir when taking heat converted into work, (ii) limitations in this conversion of heat to work, (iii) the impossibility of cooling down the coldest body by means of the work performed by this body? Thus, it can be said that Kelvin did not provide his own formulation of the II law of thermodynamics in a very strict and careful manner [2]. Therefore, it is not surprising that we use a version of the rule of the type K slightly different from the original formulation (see also [33]). Nevertheless, Kelvin's formulation of the II law of thermodynamics is dated to 1851 in connection with his lectures published in the form of articles in two journals [24,25]. In addition, in the second journal, Kelvin separated a short article in which he repeated the formulation of his principle, and also gave a general interpretation of the II law of thermodynamics [32].

It can be concluded that the Kelvin principle in the K version (not original Kelvin version) is the same as the Ostwald principle, which says that there is no *perpetuum mobile* of type II (π_{II}) [26,28]. In other words, there is no heat engine with an efficiency of 100%:

$$\eta = \frac{W}{Q_{in}} = \frac{W}{Q_1} < 1. \tag{5}$$

It is assumed here that the heat source has a higher temperature than the heat receiver ($T_{in} = T_1 > T_{out} = T_2$). Formally, however, it is possible to consider taking heat from a reservoir with a lower temperature (conventionally $Q_2 < 0$) and converting it into work $W > 0$:

$$\eta_2 = \frac{W}{\mid Q_2 \mid} < 1. \tag{6}$$

And for the processes going in the opposite direction, the equivalent of this efficiency will have the form $\tilde{\eta}_2 = \mid W \mid / \mid Q_2 \mid$.

At first glance, the Kelvin principle seems to be very trivial and seems to be significantly weaker than the Carnot principle. Nevertheless, it is allegedly proved that the Kelvin principle in the form of K is equivalent to the full formulation of the II law of thermodynamics [12–15]. Criticism of this proof based on combining engine and heat pump processes could only be found in the aforementioned work by a scientist Bhattacharyya [16].

Clausius II principle (CII)
There is a function of the state of the thermodynamic system called entropy, which the dependence of the value on time allows to determine the system's following of the thermodynamic equilibrium. Namely, in thermally insulated systems, the only possible processes are those in which the total entropy of the system (two heat reservoirs) does not decrease:

$$\Delta S = \Delta S_1 + \Delta S_2 \geq 0. \tag{7}$$

For reversible processes, the increase in total entropy is zero, and for irreversible processes it is greater than zero. The change in entropy of a reservoir is here understood as the ratio of the heat introduced into it to its temperature.

Thus, when using standard notations, Clausius II principle takes the following form for the processes considered here:

$$\Delta S = \frac{-Q_1}{T_1} + \frac{Q_2}{T_2} \geq 0. \tag{8}$$

It is assumed that a specific device operating cyclically between heat reservoirs does not change its entropy by definition of cyclicity. Additionally, similarly for the C0 principle, the absolute temperature scale is assumed here.

The Clausius II principle is the most elegant formulation of the II law of thermodynamics. Some textbooks only introduce an entropy formulation of this principle (e.g., [34]). Only the definition of entropy presents some difficulty and challenge here. Clausius's definition of entropy (integral of heat divided by temperature) is not always general enough in thermodynamics. However, we assume that it is sufficient for the process diagrams considered here. Another, usually overlooked, subtlety of the II law of thermodynamics (Clausius II principle and others) is that it actually speaks of states beyond the thermodynamic equilibrium ($T_1 \neq T_2$) to which the system is approaching but which it does not necessarily achieve (e.g., on diagrams with unlimited heat reservoirs—thermostats).

The year 1865 of the publication of work [35] is regarded to be the year when principle CII was created, even though in the years 1854 and 1856 in Clausisus works [19,20] there were expressions of the type $N = \sum Q/T$ or $N = \int dQ/T$ with the correct analysis of signs ($N > 0$ for irreversible processes, $N < 0$ for impossible and $N = 0$ for reversible). The use of inequalities in physics would be something completely new, so in the 1850s in Clausius works they do not appear explicitly. Inequalities in the mathematical formula can be found in Clausius, for example, in the textbook [22] from 1867. In this textbook, Clausius also changes the notation of N to S by calling this quantity a transformation value (process measure). Then, translating *transformation* into Greek ($\tau\rho o\pi\eta$), Clausius introduces the term *entropy*. Of course, Clausius distinguished the entropy change in a process from its absolute value. The concept of entropy entered the canon of physics, and the principle of its growth (more precisely, its not decreasing) constitutes, for example, the term *entropy production* [36]. This form of the principle is valid in elementary systems of statistical mechanics [37].

There are also other attempts to formulate the II law of thermodynamics. The formulation by Caratheodory is sometimes given: "*In the surroundings of each thermodynamic state, there are states that cannot be achieved by adiabatic processes*". However, such a formulation is expressed in a completely different language than that considered in this work. Moreover, this statement sounds as tautological as the statement: *there are places in the mountains that are inaccessible to hiking trails*. At the same time, Caratheodory's statement is given very often without strict proof of equivalence with other formulations (see [29,31]). From a contemporary critique of Caratheodory's formulation in preprint [38] we also learn that this formulation was greatly criticized by Planck.

The second law of thermodynamics is also formulated in the language of statistical physics. The outstanding physicist Ludwig Boltzmann was the first to undertake this task. Unfortunately, this article is limited to phenomenological thermodynamics only.

In the context of thermodynamic considerations, the abstract concept of *perpetuum mobile* often appears. It is, therefore, worthwhile to clarify and differentiate this concept as follows.

***Perpetuum mobile* ($\pi_0, \pi_I, \pi_{II}, \pi_{III}$)**
A perpetuum mobile is a hypothetical machine, the movement and efficiency of which would contradict certain recognized laws of physics. The most faithful meaning of the Latin name ("forever moving") perpetuum mobile type 0 (π_0) would not be stopped in spite of the existence of frictional forces and resistance to motion. A perpetuum mobile type I (π_I) would do work from nothing (or almost nothing—efficiency greater than 100%) against the principle of conservation of energy and against the I law of thermodynamics. A perpetuum mobile type II (π_{II}) would be 100% efficient and would be against the Kelvin principle. On the other hand, the perpetuum mobile type III (π_{III}) would have efficiency lower than 100%, but higher than the maximum efficiency predicted by the Carnot principle.

Most often it only stands out *perpetuum mobile* π_I and π_{II}. However, *perpetuum mobile* π_{III} is crucial for this article. Sometimes the term *perpetuum mobile* of type III is understood as π_0. However, the proposed terminology seems clear and consistent with the meaning of the Latin words *perpetuum mobile* for π_0 and consistent with the classification of various ways of doing work by $\pi_I, \pi_{II}, \pi_{III}$—while machine π_0 does not do the work. A similar notation to *perpetuum motion* will also be applied to the $\pi^{\uparrow}$ process where heat spontaneously flows uphill (towards higher temperature). Such a process, despite its similarities, does not have to be automatically equivalent to π_{II}.

Laws of physics, such as the thermodynamic principles under consideration, are subject to some kind of experimental verification. Unfortunately, from a formal point of view, these principles are not subject to truth proof like mathematical theorems. Nevertheless, the two principles may result from each other. Unfortunately, the implication of principles is something more complex than the material implication of Table 1, that depends only on logical values.

Table 1. Table of logical values of material implication and (material) equivalence.

p	q	$p \Rightarrow q$	$p \Leftrightarrow q$
0	0	1	1
0	1	1	0
1	0	0	0
1	1	1	1

Two principles may be subject to a relationship of formal implication. The formal implication related to a specific theoretical system implemented by certain models becomes a semantic implication. A provable formal implication is called syntactic implication. Kurt Gödel proved the theorem that a semantic implication is a syntactic implication. In other words, a model-based system is needed to prove formal implication. In the case of thermodynamic principles, models are sets of processes (diagrams) that follow a given principle. Thus, the formal implication of the two principles can be written as follows:

$$Z_1 \mid\!\rightarrow Z_2 \Leftrightarrow \forall d \in \Omega: \ d \in \{Z_1\} \Rightarrow d \in \{Z_2\}, \tag{9}$$

where $\{Z\}$ stands for the model, i.e., the set of processes consistent with the Z principle. If there are several (or many) models, the formal implication should apply for all Z_1 models, and there should be at least one Z_2 model containing the same d process. The symbol of formal (syntactic) implication $\mid\!\rightarrow$ used here was created from the combination of the symbol known from the formal logic $\vdash$ and the usual symbol of the implication arrow $\rightarrow$ (see [39]). Strictly speaking, this symbol in LaTeX transcription is a combination of the sign $|$ with the sign $\rightarrow$ using a negative space ("\med \!\rightarrow"). In formal logic, the symbol $\models$ is also used for the semantic formal implication. In model theory, this symbol is synonymous, but it is used to constitute a sentence, thesis, or a principle within some system [40]. For example, the fact that in the system of theoretical thermodynamics $\mathcal{T}$ the Kelvin law is true we write as $\mathcal{T} \models K$ or more precisely $\mathcal{T} \models (CII \mapsto K)$, where we assume the truth of the CII principle.

We can see that the system of models (sets of processes) allows us to reduce the formal implications to the material implications. The significant difference, however, is that the material implication must hold true for all the processes (diagrams) of the system to which the formal implication relates. This also applies to transposition law, which is used in the scheme of proof by contradiction:

$$Z_1 \mapsto Z_2 \Leftrightarrow \forall d \in \Omega: \; d \notin \{Z_2\} \Rightarrow d \notin \{Z_1\}. \tag{10}$$

Thus, in order to prove the implication of the principles, it is not enough to point to the d' process, which contradict the Z_2 principle, and to show that it implies a contradiction with the Z_1 principle (compare Figure 2a with Figure 2b). It is also worth noting that proof by contradiction is not the most valued method of proof. Constructive proof that follows the original direction of implication is the most valued. There is even an orthodox version of logic (intuitionistic logic), that rejects the law of the excluded middle—including proof by contradiction. There have even been attempts to integrate intuitionism into the mathematical proving system. The Brouwer's program is the best known in this respect.

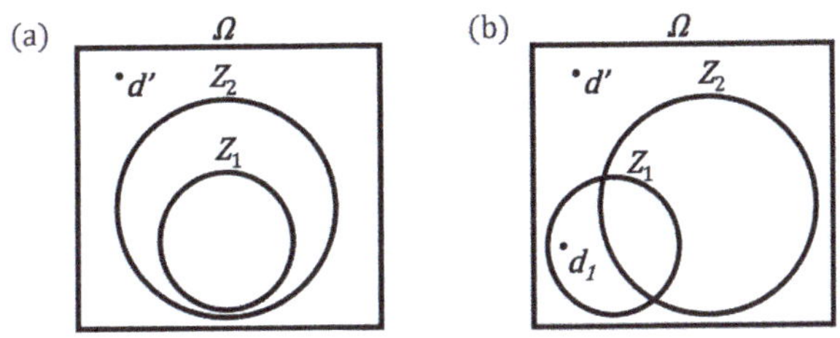

Figure 2. Schematic representation of the relationship between two physical principles Z_1 and Z_2 in a set of virtual physical processes Ω ($\{\}$ brackets to distinguish the model from the principle are omitted here): (**a**) The case of implication of principles means the inclusion of appropriate sets (models). (**b**) A case of principles that do not implicate each other. The counterexample in the form of the d_1 process disproves the implication $Z_1 \mapsto Z_2$, despite pointing to the d' process that would allegedly prove this implication by contradiction.

So, in order to prove the $Z_1 \mapsto Z_2$ implication of the principles, it is necessary to show that the sets of processes conforming to these principles (their models) are subsets $\{Z_1\} \subset \{Z_2\}$. The set of the resulting (often weaker) principle should, therefore, be a set greater than or equal to the set of the previous (often stronger) principle. It is a little easier to rebut the implications of two principles. For this purpose, it is enough to indicate the d_1 process, which belongs to the $\{Z_1\}$ set, and does not belong to the $\{Z_2\}$ set (see Figure 2b).

The principles are formally equivalent when the formal implication acts both ways:

$$Z_1 \leftrightarrow Z_2 \Leftrightarrow (Z_1 \mapsto Z_2) \wedge (Z_2 \mapsto Z_1) \Leftrightarrow \forall d \in \Omega: \; (d \in \{Z_1\} \Leftrightarrow d \in \{Z_2\}). \tag{11}$$

Usually the equivalence of the three principles is proved by a looped chain of implications $Z_1 \mapsto Z_2 \mapsto Z_3 \mapsto Z_1$ (or just as well in the opposite direction). However, this way of proving is prone to error—it is enough to undermine the weakest link in the chain loop for the equivalence proof to be invalid. A more comprehensive approach is to analyze the entire matrix of implications (Table 2).

Table 2. Full *status quo* matrix of the mutual implications of the four principles related to the II law of thermodynamics. Question mark "?" refers to the vague *status quo* of implication, and the annotation "(?)" refers to the implications questioned by the author.

Principle	Carnot	Clausius I	Kelvin	Clausius II
Carnot	$\equiv$	$C0 \mapsto CI$?	$C0 \mapsto K$?	$C0 \mapsto CII$?
Clausius I	$CI \mapsto C0$ (?)	$\equiv$	$CI \mapsto K$ (?)	$CI \mapsto CII$ (?)
Kelvin	$K \mapsto C0$ (?)	$K \mapsto CI$ (?)	$\equiv$	$K \mapsto CII$ (?)
Clausius II	$CII \mapsto C0$	$CII \mapsto CI$	$CII \mapsto K$	$\equiv$

We can see that of the 12 implications, 3 do not have a clear *status quo*. The point is that Carnot's principle is not given as an equivalent formulation of the II law of thermodynamics, but as a consequence of this law. Moreover, we see that as many as 6 other implications are being questioned. The point is that the author supposes that the principles of Clausius I and Kelvin are not equivalent to each other and are not equivalent to the II law of thermodynamics. The main goal of this paper is to formally prove these hypotheses.

2. Materials and Methods

The work is theoretical in nature, based on strict proof of the formal implications of the relevant principles and the rebuttals of the remaining formal implications, which are not satisfied. The considered formal implications relate to the defined simplified conceptual system of thermodynamics. This system consists of a set of Ω of all real or virtual processes (diagrams), along with some elementary operating rules for these processes.

An exemplary engine diagram is shown in Figure 3. For convenience, such a diagram will be marked with the arrows $\downarrow\rightarrow\downarrow$, informing respectively that: heat flows from the heater, work is performed by the device (engine), heat flows to the cooler. The arrows of such a diagram determine the signs of heat and work in the associated record of the (Q_1, W, Q_2) process, which in this case of the engine process means the positivity of all parameters: $Q_1 > 0, W > 0, Q_2 > 0$. If an arrow is pointing in the opposite direction, it means that the part of the process is going the opposite way and the corresponding heat or work symbol is negative (Figure 4). However, if a given part of the process does not occur, it is marked with 0, e.g., diagram $\downarrow 0 \downarrow$ represents a process in which heat flows from the heater to the cooler without any work being done. It is assumed that the symbols of the diagrams represent (apart from the null 000 process) a continuum of processes with different values of heat and work following the arrows (and the I law of thermodynamics). In addition, diagrams specifying the efficiency of the engine $(\downarrow\rightarrow\downarrow)_\eta$ or the efficiency of the heat pump $(\uparrow\leftarrow\uparrow)_{\bar{\eta}}$ will also be considered.

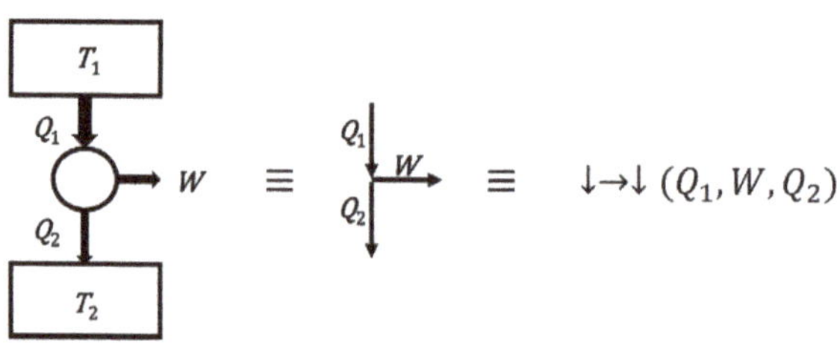

Figure 3. Schematic drawing of the engine process with the equivalent arrows diagram and the equivalent notation used in the text. Most often, the engine process denoted by capital letters will denote a Carnot efficient engine. The $\downarrow\rightarrow\downarrow$ engine process with a different efficiency may then be labeled as $(Q_1 \pm q, W, Q_2 \pm q)$ or (q_1, w, q_2).

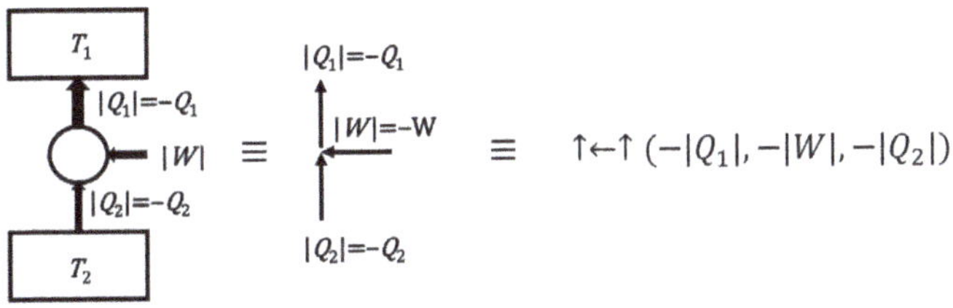

Figure 4. Schematic drawing of the cooler process with the equivalent arrows diagram and the equivalent notation used in the text. The directions of the arrows show the actual direction of heat flow or how the work was done. The heat and work symbols refer nominally to the engine process, so they have negative values here.

As mentioned before, *a priori* is assumed that the diagrams satisfy the law of conservation of energy, or the I law of thermodynamics: $Q_1 = W + Q_2$. Additionally, in the diagram set, a rule for adding diagrams is introduced, which physically means linking the processes. Adding diagrams is used, among other things, to define the concept of a convex of a set of diagrams:

The definition of a convexity (for a set of diagrams—for a model)
A set of diagrams will be called convex if the sum of any two diagrams of this set belongs to this set. In other words, the convexity of the set requires that the add diagrams operation be internal to that set.

Thus, the definition of the convexity of the set is here simplified to the operation of adding elements without having to consider multiplying diagrams by non-negative real numbers. However, the ability to scale diagrams means that the above definition is equivalent to the standard geometric definition of a convex set. The definition of a convex set helps to formulate the following important condition:

Completeness condition (for model of principle)
It is assumed that the set of diagrams (processes), consistent with a given principle and called the model of this principle, must be the largest convex set possible. It is allowed to have many alternative sets (models) that meet the above condition, in the sense that a given set (model) cannot be enlarged without breaking compliance with the principle or without breaking the convexity requirement.

The completeness condition allows for the consideration of richer models (sets of diagrams) for a given principle. Thanks to this condition, diagrams (processes) whose compliance with the principle is not obvious can be attached to the model. An attached diagram (process) can be created as a result of adding two trivial diagrams or diagrams already belonging to the model. In addition, a diagram (process) can be attached to the model, the addition of which to existing diagrams results in a trivial diagram or a diagram belonging to the model. This procedure can fork the model into two models or the entire model family. Regardless of the number of models in the family, each model should be considered an alternative independent model. If there is only one model, it will be called a homogeneous model here, and in the literature it is sometimes called a stable model.

Given the above theoretical structure, we can study the formal implications (or lack thereof) between the four elementary principles of thermodynamics $(C0, CI, K, CII)$. There are as many as 12 formal implications from Table 2 to be proved or disproved. Proof of the $Z_1 \mapsto Z_2$ implication will be performed double, once by checking the content of the sets of diagrams (models) of both principles $\{Z_1\} \subset \{Z_2\}$, two by formal constructive proof (proof by contradiction will not be used). Similarly, disproving the implications will consist, firstly, in checking that the models do not contain themselves $\{Z_1\} \not\subset \{Z_2\}$, and secondly, in pointing to a counterexample in the form of a diagram d_1 consistent with principle Z_1, but inconsistent with principle Z_2 (see Figure 2b). The negation of the formal implication will be denoted by the standard negation operator $\neg(Z_1 \mapsto Z_2)$ or by the crossed arrow $Z_1 \not\mapsto Z_2$.

The proof and rebuttals will be presented in the Discussion section and summarized in the Results section. A synthesis will be given in the Conclusions.

3. Discussion: Principles Structure, Proof and Rebuttals

Before starting the analysis of formal implications between thermodynamic principles, it is worth writing out sets of diagrams (models) for individual principles. Some principles have a complex structure of models that take into account different possible scenarios for realizing a given principle.

3.1. Structure of the Carnot Principle Model

The main processes mentioned in the Carnot principle are $\downarrow\rightarrow\downarrow$ engine processes with efficiency not higher than η_C. This set can be extended to include limit cases in the form of processes: zero efficiency $\downarrow 0 \downarrow$ and null process 000.

The remainder of the model will be found by eliminating processes, that are inconsistent with the condition of internality of adding processes. Consider the $(\downarrow\rightarrow\downarrow)_{\eta_C}$ engine process with a maximum Carnot efficiency of η_C. Let us describe the parameters of this non-zero process with three values (Q_1, W, Q_2). Less efficient processes have the form $(Q_1 + q, W, Q_2 + q)$, where $q > 0$. Note that such processes are the sum of the Carnot process with the maximum efficiency $(\downarrow\rightarrow\downarrow)_{\eta_C}$ and the process with zero efficiency $\downarrow 0 \downarrow$ $(q, 0, q)$. On the other hand, adding the $\uparrow 0 \uparrow$ $(-q, 0, -q)$ process to the Carnot process would lead to obtaining a virtual (impossible) process with efficiency greater than Carnot efficiency. Thus, processes of type $\uparrow 0 \uparrow$ should be excluded from the Carnot principle model. Similarly, we will exclude type $\uparrow\leftarrow\uparrow$ $(-Q_1 - q, -W, -Q_2 - q)$ refrigeration processes, which are the inverse of engine processes with less than maximum efficiency. Well, by adding the Carnot process to this process, we get a process outside the model ($q > 0$):

$$(-Q_1 - q, -W, -Q_2 - q) + (Q_1, W, Q_2) = (-q, 0, -q) \notin \{C0\}. \tag{12}$$

In addition to the excluded refrigeration processes, there are form $\uparrow\leftarrow\uparrow$ $(-Q_1 + p, -W, -Q_2 + p)$ refrigeration processes which are not excluded for $p \geq 0$. It is most convenient to assign the efficiency parameter $\tilde{\eta}$ to such refrigeration processes, which refers to the inverse engine process. Although the inverse engine process is virtual ($\tilde{\eta} \geq \eta_C$), the refrigeration process $(\uparrow\leftarrow\uparrow)_{\tilde{\eta}}$ exists and belongs to the Carnot principle model.

A borderline case of the ineffective refrigeration process of $\uparrow\leftarrow\uparrow$ for $\tilde{\eta} \rightarrow 1$ is the process of converting work into heat of the heater $\uparrow\leftarrow 0$. The superposition of the $\uparrow\leftarrow 0$ process with an appropriately selected $\downarrow 0 \downarrow$ process gives the $0 \leftarrow\downarrow$ process. And combining the last two processes results in a new type of process $\downarrow\leftarrow\downarrow$. However, as we add a small heat flow in the form $\downarrow 0 \downarrow$ to the $\uparrow\leftarrow 0$ process, we get $\uparrow\leftarrow\downarrow$.

In summary, the Carnot principle model consists of the following processes:

$$\{C0\} = \{000,\ \downarrow 0 \downarrow,\ 0 \leftarrow\downarrow,\ \uparrow\leftarrow 0,\ \downarrow\leftarrow\downarrow,\ \uparrow\leftarrow\downarrow,\ (\downarrow\rightarrow\downarrow)_{0<\eta\leq\eta_C}, (\uparrow\leftarrow\uparrow)_{\eta_C\leq\tilde{\eta}<1}\}. \tag{13}$$

We can see that the uniform (stable) model of Carnot principle consists of diagrams of 8 types. There are $3 \times 3 \times 3 = 27$ possible all diagram types (make sense or not). However, some diagrams contradict the principle of conservation of energy: with two zeros $3 \times 2 = 6$, with one zero $3 \times 2 = 6$, without zeros 2—which adds up to $6 + 6 + 2 = 14$ for diagrams outside areas of consideration. So in the domain of consideration there are $27 - 14 = 13$ of diagram types. This means that the Carnot principle model rejects five types of diagrams.

Therefore, it is worth listing the complement to the Carnot model (a set of impossible processes).

$$\overline{\{C0\}} = \{\uparrow 0 \uparrow,\ 0 \rightarrow\uparrow,\ \downarrow\rightarrow 0,\ \uparrow\rightarrow\uparrow,\ \downarrow\rightarrow\uparrow,\ (\downarrow\rightarrow\downarrow)_{\eta_C<\eta<1}, (\uparrow\leftarrow\uparrow)_{0<\tilde{\eta}<\eta_C}\}. \tag{14}$$

Indeed, five new types of diagrams were created simply by changing the direction of the corresponding arrows. The other two diagrams do not differ in type but have different ranges of efficiency values. Thus, combinatorics is here in line with the resulting model of Carnot's principle.

3.2. Structure of Models for Clausius I Principle

The Clausius principle I has a whole family of models, that can be parameterized with the maximum efficiency $\eta_m \in [0, 1]$:

$$\{CI\} = \{CI\}_{\eta_m} = \{CI\}_0, \{CI\}_{0<\eta_m<1}, \{CI\}_1. \quad (15)$$

We have a separate principle model for each value of maximum efficiency. The principle allows for both a zero efficiency model containing only refrigeration processes without engine processes, and also allows for an efficiency model of 1 (*perpetuum mobile* π_{II}) without refrigeration processes:

$$\{CI\}_0 = \{000, \downarrow 0 \downarrow, 0 \leftarrow\downarrow, \uparrow\leftarrow 0, \downarrow\leftarrow\downarrow, \uparrow\leftarrow\downarrow, no\ engines, \uparrow\leftarrow\uparrow\}, \quad (16)$$

$$\{CI\}_{0<\eta_m<1} = \{000, \downarrow 0 \downarrow, 0 \leftarrow\downarrow, \uparrow\leftarrow 0, \downarrow\leftarrow\downarrow, \uparrow\leftarrow\downarrow, (\downarrow\rightarrow\downarrow)_{0<\eta\leq\eta_m}, (\uparrow\leftarrow\uparrow)_{\eta_m\leq\bar{\eta}<1}\}, \quad (17)$$

$$\{CI\}_1 = \{000, \downarrow 0 \downarrow, 0 \leftarrow\downarrow, \uparrow\leftarrow 0, \downarrow\leftarrow\downarrow, \uparrow\leftarrow\downarrow, no\ refrigerators, \downarrow\rightarrow 0, \downarrow\rightarrow\downarrow\}. \quad (18)$$

The first six trivial processes are common to all models, but the engine and refrigeration processes already differ in the sense that in extreme cases one of them does not occur.

3.3. Structure of Models for Kelvin Principle

The structure of the Kelvin principle models is the most complex of all the principles considered in this paper. It turns out that this principle allows scenarios of models with both "top-down" $\downarrow$ and "bottom-up" $\uparrow$ heat flows. The zero-efficiency scenario model allows for both directions simultaneously. Thus, the structure of the Kelvin models can be represented as follows:

$$\{K\} = \{K\}_0^{\updownarrow}, \{K\}_{0<\eta_m<1}^{\downarrow}, \{K\}_{1^-}^{\downarrow}, \{K\}_{0<\eta_m<1}^{\uparrow}, \{K\}_{1^-}^{\uparrow}. \quad (19)$$

Models containing processes of arbitrarily high efficiency but less than one (1^-), which should not be confused with models with a specified maximum efficiency of η_m, have been distinguished here. The zero-efficiency model is as follows:

$$\{K\}_0^{\updownarrow} = \{000, \downarrow 0 \downarrow, 0 \leftarrow\downarrow, \uparrow\leftarrow 0, \downarrow\leftarrow\downarrow, \uparrow\leftarrow\downarrow, no\ engines, \uparrow\leftarrow\uparrow, \uparrow 0 \uparrow\}. \quad (20)$$

Models with a natural direction of heat flow ("downwards") and with an intermediate maximum efficiency coincide with the analogous models for the Clausius I principle:

$$\{K\}_{0<\eta_m<1}^{\downarrow} = \{000, \downarrow 0 \downarrow, 0 \leftarrow\downarrow, \uparrow\leftarrow 0, \downarrow\leftarrow\downarrow, \uparrow\leftarrow\downarrow, (\downarrow\rightarrow\downarrow)_{0<\eta\leq\eta_m}, (\uparrow\leftarrow\uparrow)_{\eta_m\leq\bar{\eta}<1}\}. \quad (21)$$

On the other hand, the model containing the efficiencies arbitrarily close to one, differs from the analogous model of the CI principle only by the lack of processes with the efficiency of one:

$$\{K\}_{1^-}^{\downarrow} = \{000, \downarrow 0 \downarrow, 0 \leftarrow\downarrow, \uparrow\leftarrow 0, \downarrow\leftarrow\downarrow, \uparrow\leftarrow\downarrow, no\ refrigerators, \downarrow\rightarrow\downarrow\}. \quad (22)$$

In models with the opposite direction of heat flow, it is enough to reverse the directions of the corresponding arrows:

$$\{K\}_{0<\eta_m<1}^{\uparrow} = \{000, \uparrow 0 \uparrow, 0 \leftarrow\downarrow, \uparrow\leftarrow 0, \uparrow\leftarrow\uparrow, \downarrow\leftarrow\uparrow, (\uparrow\rightarrow\uparrow)_{0<\eta_2\leq\eta_m}, (\downarrow\leftarrow\downarrow)_{\eta_m\leq\bar{\eta}_2<1}\}, \quad (23)$$

$$\{K\}_{1^-}^{\uparrow} = \{000, \uparrow 0 \uparrow, 0 \leftarrow\downarrow, \uparrow\leftarrow 0, \uparrow\leftarrow\uparrow, \downarrow\leftarrow\uparrow, no\ "refrigerators", \uparrow\rightarrow\uparrow\}. \quad (24)$$

The term *"refrigerators"* is given here in quotation marks, because the heat in such a device would be pumped from a higher temperature to a lower temperature, in these models. It is worth noting that these models do not have the usual engines (*no engines*), i.e., processes in which the work is done at the expense of the heat of the heater, not the cooler. The lack of typical engines ensures the consistency of these models. Such a class of models obviously contradicts the tendency towards thermodynamic equilibrium, but the Kelvin principle does not seem to determine this condition.

Above all, however, both classes of heat flow direction, similarly to the Clausius I principle, have the efficiency values η_m, which may exceed the Carnot efficiency η_C ($\eta_m > \eta_C$). The value of $\eta_m < 1$ is not related in any way to the value of η_C. Hence, we have whole classes (sets families) of Kelvin (and Clausius I) models. Models for which $\eta_m > \eta_C$ contain more engine diagrams than the uniform (stable) and common model for the Carnot principle and Claussius II principle.

3.4. Structure of the Clausius II Principle Model

The principle CII about non-decreasing entropy of physical processes is consistent with adding processes and the condition of completeness of the model. In other words, the compliance of processes with the CII principle automatically guarantees the internality of adding processes. As a result, the model of the Clausius II principle is a homogeneous single set (stable model):

$$\{CII\} = \{000, \downarrow 0 \downarrow, 0 \leftarrow\downarrow, \uparrow\leftarrow 0, \downarrow\leftarrow\downarrow, \uparrow\leftarrow\downarrow, (\downarrow\rightarrow\downarrow)_{0<\eta\leq\eta_C}, (\uparrow\leftarrow\uparrow)_{\eta_C\leq\bar{\eta}\leq 1}\}. \quad (25)$$

The above model closely follows the Carnot principle model:

$$\{CII\} = \{C0\}, \quad (26)$$

but it is narrower than the families of models for Clausius I and Kelvin principles. Strictly speaking, the $\{CII\}$ model coincides with only one model $\{CI\}_{\eta_m=\eta_C}$ of Clausius I principle and only one model $\{K\}^{\downarrow}_{\eta_m=\eta_C}$ of Kelvin principle. Therefore, Clausius II principle and the equivalent Carnot principle are stronger principles than those of Clausius I and Kelvin.

The structure of the (13), (15), (19), (25) models already specifies the table of all implications or their absence. However, these implications will be subject to a more detailed proof analysis.

First, the implications of the remaining principles from the Carnot principle will be analyzed.

3.5. Theorem $C0 \mapsto CI$

Proof. Based on (13) and (15), (17) we can see that $\{C0\} \subset \{CI\}_{\eta_m}$ for $\eta_m = \eta_C$. Consider a Carnot heat engine denoted by the diagram $(\downarrow\rightarrow\downarrow)_{\eta_C}$, in which heat input, work done and output heat can be written as (Q_1, W, Q_2). Similarly, we will write the process of heat flow $(q, 0, q)$, which is a flow down (towards a lower temperature) $\downarrow 0 \downarrow$ for $q > 0$ or a flow up (towards a higher temperature) $\uparrow 0 \uparrow$ for $q < 0$. Now let us combine these two processes:

$$(Q_1, W, Q_2) + (q, 0, q) = (Q_1 + q, W, Q_2 + q). \quad (27)$$

Let us only consider the resultant processes which follow the Carnot principle:

$$\eta = \frac{W}{Q_1 + q} \leq \frac{W}{Q_1} = \eta_C \quad \Rightarrow \quad q \geq 0. \quad (28)$$

Therefore, on the basis of the completeness condition, we can conclude that the Carnot principle is compatible with the processes of heat flow downwards $\downarrow 0 \downarrow$ ($q > 0$), and the processes of heat flow upwards $\uparrow 0 \uparrow$ ($q < 0$) do not occur. Thus, we obtained the thesis of Clausius I principle. □

3.6. Theorem $C0 \mapsto K$

Proof. Based on (13), (19) and (21) we can see that $\{C0\} \subset \{K\}^{\downarrow}_{\eta_m}$ for $\eta_m = \eta_C$. Since essentially $\{K\} = \{K\}^{\updownarrow}_{0}, \{K\}^{\downarrow}, \{K\}^{\uparrow}$, it can be assumed that $\{C0\} \subset \{K\}$. Thus, the set of Carnot principle processes follows the Kelvin principle. These are, of course, processes with the natural direction of heat flow downwards (as in principle of Clausius I).

Consider a process involving an $\downarrow\rightarrow?$ heat engine in which we do not know if the cooler absorbs heat ($Q_2 =?$). Let us describe this process with three parameters (Q_1, W, Q_2), where $Q_1 > 0, W > 0, Q_2 = Q_1 - W$. The Carnot principle implies the condition for Q_2:

$$\eta = 1 - \frac{Q_2}{Q_1} \leq \eta_C < 1 \quad \Rightarrow \quad Q_2 \geq (1 - \eta_C)Q_1 > 0. \tag{29}$$

This means that heat must be transferred to the cooler, so it is a $\downarrow\rightarrow\downarrow$ process with less than 1 efficiency, which is in line with the Kelvin principle. □

3.7. Theorem $C0 \mapsto CII$

Proof. Based on (13) and (25) we can see that $\{C0\} = \{CII\}$, so in particular also $\{C0\} \subset \{CII\}$. Thus, on the basis of the models (sets of processes), there is implication (or even equivalence).

Consider the Carnot engine process $\downarrow\rightarrow\downarrow$ (Q_1, W, Q_2) and the natural heat flow downward process $\downarrow 0 \downarrow$ $(q, 0, q)$. On the basis of (28) within the Carnot principle we know that the process inverse to the second process cannot take place (so $q \geq 0$). The completeness condition implies the existence of a summary process $(Q_1 + q, W, Q_2 + q)$. Let us calculate the entropy change of the heat reservoirs in this process:

$$\Delta S = \frac{-Q_1 - q}{T_1} + \frac{Q_2 + q}{T_2} = 0 + \frac{T_1 - T_2}{T_1 T_2} q \geq 0. \tag{30}$$

which is in line with the Claussius II principle.

Now consider the Carnot refrigeration process $\uparrow\leftarrow\uparrow$ $(-Q_1, -W, -Q_2)$. By adding the $\downarrow 0 \downarrow$ $(q, 0, q)$ process to it, we get the allowed $(-Q_1 + q, -W, -Q_2 + q)$ process. The change in the entropy of the heater and cooler in this refrigeration process is:

$$\Delta S = \frac{Q_1 - q}{T_1} + \frac{-Q_2 + q}{T_2} = 0 + \frac{T_1 - T_2}{T_1 T_2} q \geq 0. \tag{31}$$

which, again, is in line with the Clausius II principle. The change of entropy would be negative for the $(-Q_1 - q, -W, -Q_2 - q)$ refrigeration process—however, such a process is not compatible with the Carnot principle, as the convexity condition allows for addition, not subtraction, of process diagrams.

Equality of process sets of the $C0$ and CII principles and the fulfillment of the CII principle for processes permitted by the $C0$ principle ends the proof. □

Now we turn to the implications of Clausius I principle. The first implication, however, does not hold true (as do the other two).

3.8. Theorem $\neg(CI \mapsto C0)$

Proof. Based on (13) and (15) we can see that the model family for the CI principle is larger than the uniform set of processes (stable model) for the $C0$ principle. These are additional models that contain engine processes with an efficiency greater than that of the Carnot cycle ($\eta_C < \eta_m < 1$). In addition, there is even a $\{CI\}_1$ model containing *perpetuum mobile* type II, which is contrary to the Carnot principle.

First of all, it should be shown that processes with the maximum efficiency in the range $\eta_C < \eta_m < 1$ are allowed by the CI principle. Although this principle distinguishes between higher and lower temperatures and the direction of heat flow, it does not impose quantitative restrictions on efficiency (even a model with $\eta_m = 1$ is possible).

Consider the Carnot engine cycle process $\downarrow\rightarrow\downarrow$ (Q_1, W, Q_2). We assume that this process complies with the Clausius I principle. Let us consider whether there is a possible model of the CI principle, in which there is a process with even greater efficiency, i.e., process $\downarrow\rightarrow\downarrow$ $(Q_1 - q, W, Q_2 - q)$, where $q > 0$. The inclusion of such a process will not lead to a contradiction, if only some refrigeration processes exist in the model. For example, $\uparrow\leftarrow\uparrow$ $(-Q_1 + q + p, -W, -Q_2 + q + p)$ refrigeration processes will be acceptable for $p \geq 0$. Their addition to the Carnot cycle process and their addition to the more efficient process under consideration leads to downward heat flow processes $\downarrow 0 \downarrow$ or to a null process 000, which are in accordance with the CI principle. In the considered model of the CI principle, however, the refrigeration process type $\uparrow\leftarrow\uparrow$ $(-Q_1, -W, -Q_2)$ cannot occur because the higher efficiency engine added to the process would generate a heat flow upwards $\uparrow 0 \uparrow$ contrary to the CI principle. However, the absence of $(-Q_1, -W, -Q_2)$ in the model does not make it contradictory. Thus, the considered model of the CI rule is acceptable. At the same time, this model of the CI principle is not compatible with the model of the C0 principle, as it contains engine processes with efficiency greater than the Carnot cycle. □

3.9. Theorem $\neg(CI \mapsto K)$

Proof. Based on (15), (18) and (19), (22) we can see that $\{CI\} \not\subset \{K\}$. The absence of inclusions here is quite subtle and is due to the lack of inclusions of special models $\{CI\}_1 \not\subset \{K\}^{\downarrow}_{1-}$. This is due to the existence of a *perpetuum mobile* process of the II type $\downarrow\rightarrow 0 \in \{CI\}_1$, which is not allowed by the Kelvin principle $\downarrow\rightarrow 0 \notin \{K\}^{\downarrow}_{1-}$.

Thus, the rebuttal of the implication in this case is based on the existence of one counterexample: the $\downarrow\rightarrow 0$ process in one model of Clausius I principle. It is about the $\{CI\}_1$ model the processes of which are listed in (18). The $\downarrow\rightarrow 0$ process in this model does not lead to a contradiction because there are no refrigeration processes in this model that could lead to a contradiction. Refrigeration processes may exist in other models, however, where there is no process $\downarrow\rightarrow 0$. The existence of the $\{CI\}_1$ model removes the formal implication between the CI principle and the K principle.

It is also necessary to point out the error in the alleged contemporary proof by the contradiction of the considered implication. Consider two processes that contradict Kelvin principle: $\downarrow\rightarrow 0(W, W, 0)$ and $0 \rightarrow\uparrow (0, W, -W)$, whose efficiency is 100%. Let us assume a non-zero refrigeration process $\uparrow\leftarrow\uparrow$ $(-Q_1, -W, -Q_2)$. Combining such a refrigeration process with the above 100% efficiency processes leads to upward heat flow processes, respectively: $\uparrow 0 \uparrow$ $(-Q_2, 0, -Q_2)$ and $\uparrow 0 \uparrow$ $(-Q_1, 0, -Q_1)$. Thus, it might seem that breaking Kelvin principle entails breaking Clausius I principle (alleged proof by contradiction). However, it is not so. First, the consideration of the $\uparrow\leftarrow\uparrow$ refrigeration process is consistent with, and not in contradiction to, the Kelvin Principle—upon which proof by contradiction should be based. Not surprisingly, considering contradicting the Kelvin principle and adopting the Kelvin principle leads to a contradiction. Second, there is nothing to prove that the combination of a virtual process (outside the model) with a process within the model that leads to a process outside the model (one or the other principle). The condition for the internality of adding processes in the completeness condition applies to the situation when both the processes being added belong to the model. Third, the alleged proof by contradiction relies on a material implication for two cases, while it should be based on a formal implication based on all possible processes. □

3.10. Theorem $\neg(CI \mapsto CII)$

Proof. Based on (15) and (25) we can see that $\{CI\} \not\subset \{CII\}$. From the whole family of $\eta_m \in [0, 1]$ models for the CI principle, only the $\eta_m = \eta_C$ model coincides with the CII principle model. Even the zero-efficiency $\eta_m = 0$ model is not included in the $\{CII\}$ model because it contains too many refrigeration processes. Thus, there is not formal implication from the CI principle to the CII principle.

Consider two counterexamples of implications in the form of processes, that are allowed in the CI principle models but not in the CII principle model. The first counterexample would be *perpetuum mobile* type II $\downarrow\!\rightarrow 0$, which was already given in the previous section in the context of the CI rule. The second counterexample should be the efficient refrigeration process $(-Q_1 - q, -W, -Q_2 - q)$ related to the Carnot cycle (Q_1, W, Q_2). It is easy to verify that for $q > 0$ the refrigeration process would increase the entropy of the heat reservoirs ($\Delta S > 0$), so it does not follow the CII rule. However, this refrigeration process will follow the CI principle model, which does not include the Carnot cycle, but the less efficient cycle $(Q_1 + q, W, Q_2 + q)$. Then adding these processes does not lead to contradiction with the model for the CI principle. The indicated counterexamples disproves the implication $CI \mapsto CII$.

Errors in the alleged proof of the disproved implication will now be shown. Consider the virtual engine process $\downarrow\!\rightarrow\!\downarrow$ $(Q_1 - q, W, Q_2 - q)$ with efficiency greater than that of a Carnot engine $\downarrow\!\rightarrow\!\downarrow$ (Q_1, W, Q_2) for $q > 0$. Such a process obviously increases the entropy of the system and is against the CII principle. Additionally, consider the Carnot refrigeration cycle $\uparrow\!\leftarrow\!\uparrow$ $(-Q_1, -W, -Q_2)$. The superposition of the virtual engine cycle with the Carnot refrigeration cycle gives:

$$(Q_1 - q, W, Q_2 - q) + (-Q_1, -W, -Q_2 - q) = (-q, 0, -q), \tag{32}$$

which is contrary to the CI principle of the upward heat flow process $\uparrow 0 \uparrow$. One might think that this is proof by contradiction for the implication under consideration. However, nothing could be more wrong. Firstly, the simultaneous use of the process contrary to the CII principle and the second process consistent with the principles of CI and CII raises doubts as to whether the CII principle was actually negated (in the proof by contradiction). Secondly, the proof should apply to all processes, not just one process (material implication vs formal implication).

Thus, based on the lack of inclusion of models, two counterexamples, and an error indication in the supposed typical proof, the implication $CI \mapsto CII$ is rebutted. □

It will now be shown that the other principles do not formally follow from the Kelvin principle.

3.11. *Theorem* $\neg(K \mapsto C0)$

Proof. Based on (13) and (19) we can see that $\{K\} \not\subset \{C0\}$. The K principle has rich family of models, so they cannot be contained in a single model for the $C0$ principle. At the level of the model structure, the lack of the considered formal implication is quite clear.

However, let us give two counterexamples of processes that conform to the K rule, but not the $C0$ rule. Let it be the $\downarrow\!\rightarrow\!\downarrow$ $(Q_1 - q, W, Q_2 - q)$ engine process with any high efficiency less than 100 % ($q < Q_2, q \approx Q_2$) and the $\uparrow 0 \uparrow$ process heat flow upward (towards higher temperature). These processes are found in the models of K principle and do not appear in the model of $C0$ principle. Contrary to appearances, the inverse Carnot cycle $(-Q_1, -W, -Q_2)$ cannot be added to the engine process in order to obtain a contradictory process $\uparrow 0 \uparrow$ for these models. This cannot be done because the ultra high efficiency model does not have cooling cycles for $\tilde{\eta} = \eta_C$. However, when it comes to the $\uparrow 0 \uparrow$ process, it is possible in separate models that implement the scenario of spontaneous heat flow upward (towards a higher temperature). The Kelvin Principle does not speak of heat reservoirs temperatures, so it should come as no surprise that it determines neither the heat direction nor the Carnot efficiency. □

3.12. *Theorem* $\neg(K \mapsto CI)$

Proof. Based on (19) and (15) we can see that $\{K\} \not\subset \{CI\}$. Both principles have large families of models, but the Kelvin family of models is larger with models with opposite heat flow directions.

Counterexamples for the considered implication may be the process of heat flow upward $\uparrow 0 \uparrow$ or a process of finite efficiency $\uparrow\!\rightarrow\!\uparrow$ of heat conversion from the lower temperature reservoir to work. Both processes are simultaneously included in the two models $\{K\}^{\uparrow}_{0<\eta_2<\eta_m}$, $\{K\}^{\uparrow}_{1-}$ of Kelvin principle. Within these models, these processes cannot be excluded by adding another process from a given model, because as one can check the operation of addition, it is an internal operation in these models. At the same time, the indicated two counter-examples are directly contrary to the CI principle.

Now the errors in the supposed typical proof of $K \mapsto CI$ implication will be pointed out. This typical pseudo-proof is proof by contradiction. It relies on the $\uparrow 0 \uparrow$ $(-Q_2, 0, -Q_2)$ process, which contradicts the CI principle ($Q_2 > 0$). This process is then combined with the engine process $\downarrow\rightarrow\downarrow$ (Q_1, W, Q_2):

$$(-Q_2, 0, -Q_2) + (Q_1, W, Q_2) = (Q_1 - Q_2, W, 0). \tag{33}$$

The resulting process $\downarrow\rightarrow$ is a *perpetuum mobile* type II, which contradicts the K principle. Then it is claimed that the $K \mapsto CI$ implication has been proved, but this is not true. First, it is not clear whether it is allowed to add a process $\downarrow\rightarrow\downarrow$ that complies with the CI principle, since we want to contradict this principle in the proof by contradiction. The result of adding this process is a process that formally no longer contradicts the CI principle. Secondly, the formal implication should be checked for all processes, not just for one type of processes $\uparrow 0 \uparrow$ in the proof scheme by contradiction. There are more processes that contradict the CI principle, e.g., $\uparrow\rightarrow\uparrow, 0 \rightarrow\uparrow$.

Thus, the structure of the K and CI models and the two counterexamples, as well as the alleged proof errors, disprove the implication $K \mapsto CI$. □

3.13. *Theorem* $\neg(K \mapsto CII)$

Proof. Based on (19) oraz (25) we can see that $\{K\} \not\subset \{CII\}$. The family of models for the K principle is too large to be a subset of the CII principle model. The K principle models include too efficient engine processes $\eta > \eta_C$, as well as refrigeration processes with too low efficiency $\tilde{\eta} < \eta_C$. Such processes, on the other hand, are contrary to the CII principle.

The processes $\downarrow\rightarrow\downarrow$ $(Q_1 - q, W, Q_2 - q)$, $\uparrow\leftarrow\uparrow$ $(-Q_1 - q, -W, -Q_2 - q)$ are counterexamples for the implication under consideration. They were referenced to the $\downarrow\rightarrow\downarrow$ (Q_1, W, Q_2) process with Carnot efficiency η_C. For $q > 0$ these two counterexamples contradict the CII principle, since for them $\Delta S < 0$. However, they do not contradict the K principle, which does not allow 100% efficiency, but does not give a stronger limitation with Carnot efficiency.

However, one might get the impression that the processes $(Q_1 - q, W, Q_2 - q)$ and $(-Q_1 - q, -W, -Q_2 - q)$ do not follow the K principle because, when added to the inverse $(-Q_1, -W, -Q_2)$ or the regular (Q_1, W, Q_2) of the Carot process, they give respectively in both cases the process $\uparrow 0 \uparrow$ $(-q, 0, -q)$, which is impossible in the considered models (disregarding the existence of a separate class of models of the type $\{K\}^{\uparrow}$). The point is, however, that in the model suitable for the $(Q_1 - q, W, Q_2 - q)$ process, there is no inverse Carnot cycle with Carnot efficiency ($\tilde{\eta} = \eta_C$). For the model suitable for the $(-Q_1 - q, -W, -Q_2 - q)$ process, there is no Carnot engine cycle with Carnot efficiency ($\eta = \eta_C$). Moreover, both counterexamples of processes belong to different models, so these processes should not be added, either.

It is also worth pointing to the errors in the alleged proof by the contradiction of the implication under consideration. It is known that the $\uparrow 0 \uparrow$ $(-q, 0, -q)$ process for $q > 0$ does not comply with the CII rule because in this case $\Delta S < 0$. By combining the above process with the $\downarrow\rightarrow\downarrow$ $(q + W, W, q)$ engine process, we get *perpetuum mobile* type II, i.e., the process $\downarrow\rightarrow 0(W, W, 0)$. However, this does not prove anything, since the proof of formal implication cannot be based on one example of a process impossible for a given principle (and consequently impossible for a second principle). Moreover, considering both the impossible and the possible process (for the CII principle) is unclear. However, consider another impossible process $\downarrow\rightarrow\downarrow$ $(Q_1 - q, W, Q_2 - q)$ with an efficiency greater than that of the Carnot process. Combining it with the Carnot inverse cycle $\uparrow\leftarrow\uparrow$ $(-Q_1, -W, -Q_2)$ we get the process $\uparrow 0 \uparrow$ $(-q, 0, -q)$ impossible for the CII principle, but not excluded by the

K rule. Such a process $\uparrow 0 \uparrow$ occurs, for example, in the model $\{K\}_0^{\updownarrow}$, which includes any refrigeration process $\{K\}_0^{\updownarrow}$, but does not contain any engine process $\downarrow\rightarrow\downarrow$.

On the basis of the structure of the $\{K\}$, $\{CII\}$ models, two counterexamples and pseudo-proof errors, the formal implication of $K \mapsto CII$ has been rebutted. □

Three formal implications of the CII principle remain to be proved.

3.14. Theorem $CII \mapsto C0$

Proof. Based on (25) and (13) we can see that $\{CII\} = \{C0\}$, so in particular also $\{CII\} \subset \{C0\}$. Thus, on the basis of the models (sets of processes), the considered implication holds (and even equivalence).

However, we will show directly that the CII principle applied to the $\downarrow\rightarrow\downarrow$ (q_1, w, q_2) engine process leads to the Carnot efficiency condition:

$$\frac{-q_1}{T_1} + \frac{q_2}{T_2} \geq 0 \;\Rightarrow\; \frac{-T_2}{T_1} \geq \frac{-q_2}{q_1} \;\Rightarrow\; 1 - \frac{T_2}{T_1} \geq 1 - \frac{q_2}{q_1} \;\Rightarrow\; \eta_C \geq \eta. \tag{34}$$

Since the Carnot principle applies essentially to engine processes, the proof of formal implication may be limited here to such processes. On the other hand, the generality of the proof is additionally secured by the equality of the models. □

3.15. Theorem $CII \mapsto K$

Proof. Based on (25) and (19), (21) we can see that $\{CII\} \subset \{K\}$. There is even equality with the particular model of the result principle $\{CII\} = \{K\}^{\downarrow}_{0<\eta_m<1}$ for $\eta_m = \eta_C$.

The analyzed implication is a weaker version of the implication (34):

$$\eta \leq \eta_C = 1 - \frac{T_1}{T_2} < 1. \tag{35}$$

Thus, the K principle is weaker than the CII principle. Since the K principle applies essentially to engine processes (or the absence of *perpetuum mobile* type II), the proof can be limited to processes of this type. On the other hand, the generality of the proof is secured by the fact that one model is included in the other model. □

3.16. Theorem $CII \mapsto CI$

Proof. Based on (25) and (15), (17) we can see that $\{CII\} \subset \{CI\}$. Strictly speaking, the first-principle model is equal to the special result-principle model $\{CII\} = \{CI\}_{0<\eta_m<1}$ for $\eta_m = \eta_C$.

However, we will show directly that the principle CII applied to the processes of heat flow $(q, 0, q)$ allows heat to flow downwards $\downarrow 0 \downarrow$, but does not allow spontaneous flows upwards $\uparrow 0 \uparrow$:

$$\frac{-q}{T_1} + \frac{q}{T_2} \geq 0 \;\Rightarrow\; \frac{T_1 - T_2}{T_1 T_2} q \geq 0 \;\Rightarrow\; q \geq 0. \tag{36}$$

This condition expresses the essence of the CI principle (assuming $T_1 > T_2$), so the proof need not include other processes. On the other hand, the generality of the formal implication is secured by the inclusion of the CII principle model in the CI principle model. □

Thus, six formal implications were proved and six other formal implications were rebutted. The proof was made double: the first method was based on the inclusion relationship of models, and the second method was direct proof (not by contradiction). The rebuttals, on the other hand, were threefold: the first method showed the absence of inclusion relationship of models, the second method was based on one or two counterexamples, and the third method indicated errors in the alleged pseudo-proof by contradiction.

4. Results

The proved and disproved implications of the four principles ($C0, CI, K, CII$) are summarized in Table 3. Thus, only the two principles $C0$ and CII turned out to be equivalent. These principles are also stronger than the CI and K principles that follow from the previous ones. Thus, what Smoluchowski postulated in 1914 is being implemented: *"We call the Carnot principle as the second law of thermodynamics since Clausius time"* [41,42]. It is worth noting that the equivalence of the principles $C0$ and CII occurs in the conceptual system of thermodynamics assuming convexity and completeness of models. These additional rules were needed to give general meaning to the principles $C0$, CI, and K. The convexity assumption in the sense of internality of addition is implicitly commonly used, and the completeness rule only extends this assumption. The CII principle, on the other hand, does not need these additional rules.

Table 3. The resulting matrix of mutual formal implications (or the lack of them) of the four principles: Carnot, Clausius I, Kelvin, Clausius II. Each implication was proved or disproved, and was additionally analyzed on the basis of the structure of the models (sets of processes).

Principle	Carnot	Clausius I	Kelvin	Clausius II
Carnot	$\equiv$	$C0 \vdash\!\rightarrow CI$	$C0 \vdash\!\rightarrow K$	$C0 \vdash\!\rightarrow CII$
Clausius I	$CI \vdash\!\nrightarrow C0$	$\equiv$	$CI \vdash\!\nrightarrow K$	$CI \vdash\!\nrightarrow CII$
Kelvin	$K \vdash\!\nrightarrow C0$	$K \vdash\!\nrightarrow CI$	$\equiv$	$K \vdash\!\nrightarrow CII$
Clausius II	$CII \vdash\!\rightarrow C0$	$CII \vdash\!\rightarrow CI$	$CII \vdash\!\rightarrow K$	$\equiv$

Another way to show relationships between principles is to represent relationships between sets of all their processes belonging to all models. It has been marked, in a simplified manner, in the diagram of Figure 5. We see that the largest set of possible processes is allowed by the Kelvin principle. The only process that is not covered by the Kelvin principle, and which the Clausius I principle does not exclude, is a *perpetuum mobile* type II. Both the Clausius I principle and the Kelvin principle allow for a *perpetuum mobile* type III and, thus, differ significantly from the principles of Carnot and Clausius II. The Kelvin principle additionally includes processes of spontaneous flow of heat upwards (towards a higher temperature).

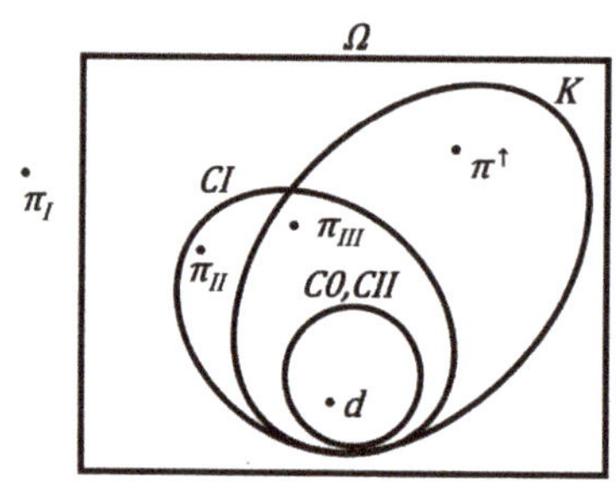

Figure 5. Resulting diagram showing the relationship between the fundamental principles crucial to the II law of thermodynamics. Additionally, the location of the fictitious *perpetuum mobile* processes of three types was marked, along with the heat flow process "upwards" $\pi^{\uparrow}$. *Perpetuum mobile* π_I goes beyond a dozen considerations, and it would be even more difficult to locate π_0 here. On the other hand, the physical process (diagram), consistent with the II law of thermodynamics (with principles $C0$ or CII), was denoted by d.

The relationships between the principles are also shown in Figure 6 in a "logical square" design. One diagonal of the square indicates equivalence, and the other diagonal is not marked, as it is neither equivalence nor implication.

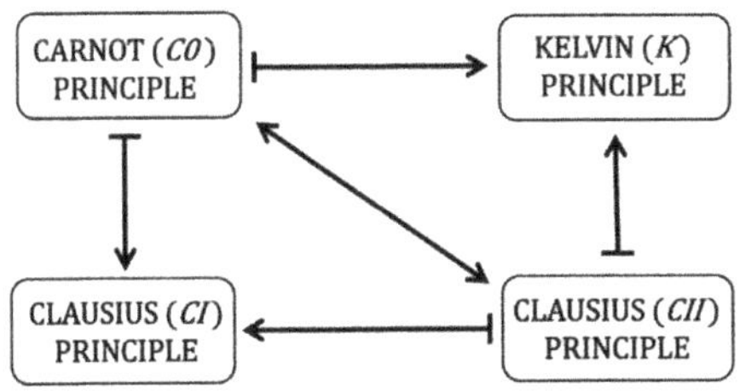

Figure 6. The resulting "logical square" of the implications of the four principles related to the II law of thermodynamics. This law is determined equivalently by the principles on the main diagonal. On the other hand, non-diagonal principles result from diagonal ones, but they are weaker and do not result from each other.

5. Conclusions

Of the four principles (Carnot, Clausius I, Kelvin, Clausius II) pretending to formulate the II law of thermodynamics, only the Carnot and Clausius II principles turned out to be equivalent and strong principles, that forbid the decrease of the entropy of a heat-insulated system. The principles of Clausius I and Kelvin turned out to be less restrictive principles that say little about the reversibility of processes, and even allow impossible processes. These principles are true, but they are also so obvious that their predictions contribute very little. Moreover, the principles of Clausius I and Kelvin turned out to be, strictly speaking, independent (they do not imply each other). However, if we omit the *perpetuum mobile* of the II kind, it can be said, that the Clausius I principle is stronger than the Kelvin principle.

The most elegant formulation of the II law of thermodynamics is the entropic principle of Clausius II. The only competing formulation of this law, as it turns out, is the equivalent Carnot principle. Carnot principle is often *implicite* treated as the equivalent of the II law of thermodynamics, but *explicite* was not included in the formulation of this principle. Instead, the formulations of the II law mistakenly included the Clausius I principle and the Kelvin principle (as well as the Caratheodory principle). It can be said that it has not been noticed that the Carnot condition $\eta \leq \eta_C$ is stronger than the Kelvin condition in terms of Ostwald $\eta < 1$. The II law of thermodynamics excludes not only the *perpetuum mobile* type II, but also the *perpetuum mobile* type III, that is a heat engine with an efficiency greater, than that of the Carnot engine. In the context of Carnot, who was a historical pioneer of the II law of thermodynamics, the distractor was the lack of knowledge of the I law of thermodynamics in his time. Carnot used the concept of indestructible heat (*caloric*), but in the sense of heat, not energy. Nevertheless, it did not prevent him from formulating correct conclusions regarding the II law of thermodynamics.

So how to explain the incorrect proof of the equivalence of Clausius I and Kelvin principles with the II law of thermodynamics? The first cause and distractor may be the frequent consideration of only Clausius I and Kelvin principles, regardless of the entropy formulation (Clausius II), or even more so the Carnot principle. In this approach, we lose the stronger context of the II law of thermodynamics. This, however, does not explain the alleged proof of a substantially incorrect implication of $K \mapsto CI$. Here a second cause appears, based on the formal implication $\mapsto$ distractor, which is a material implication $\Rightarrow$. The formal implication requires more subtle methods of proving, than the material implication. The main difference is the necessity to use the universal quantifier in formal implication, i.e., the necessity to check all the processes belonging to the model of the principle, which is the antecedent of the implication.

Funding: This research received no external funding.

Institutional Review Board Statement: Not applicable.

Informed Consent Statement: Not applicable.

Data Availability Statement: Not applicable.

Acknowledgments: Many thanks to Ludomir Newelski for his consultation on advanced logic in the context of formal implication and model theory. I also thank Martin Bier for verifying the thermodynamic aspects of the work.

Conflicts of Interest: The author declares no conflict of interest.

Abbreviations

The following abbreviations are used in this manuscript:

$C0$	Carnot principle
CI	Clausius first principle
K	Kelvin principle
CII	Clausius second principle

References

1. Carnot, S. *Réflexions sur la Puissance Motrice du Feu et sur les Machines Propres à DéVelopper Cette Puissance (Reflections on the Motive Power of Fire and on the Machines Suitable for Developing This Power)*, 1st ed.; Bachelier Libraire: Paris, France, 1824; pp. 1–58.
2. Badur, J. *Rozwój PojęCia Energii (The Development of the Concept of Energy)*, 1st ed.; Wydawnictwo Instytutu Maszyn Przepływowych PAN: Gdańsk, Poland, 2009; pp. 161–264, 726–742.
3. Thomson, W. On an Absolute Thermometric Scale founded on Carnot's Theory of the Motive Power of Heat, and calculated from Regnault's observations. *Philos. Mag. J. Sci.* **1848**, *33*, 487–490.
4. Dias P.M.C.; Pinto, S.P.; Cassiano, D.H. The Conceptual Import of Carnot's Theorem to the Discovery of the Entropy. *Arch. Hist. Exact Sci.* **1995**, *49*, 135–161. [CrossRef]
5. Thomson, W. An Account of Carnot's Theory of the Motive Power of Heat; with Numerical Results deduced from Regnault's Experiments on Steam. *Trans. R. Soc. Edinb.* **1849**, *16*, 541–574. [CrossRef]
6. Singh, N.; Sharma, U.C. *Thermodynamics as Legacy of Sadi Carnot. National Conference on Futuristics in Mechanical Engineering*; Excel India Publishers: New Delhi, India, 2015; pp. 36–42.
7. Srinivasan, J. Sadi Carnot and the Second Law of Thermodynamics. *Resonance* **2001**, *6*, 42–48. [CrossRef]
8. Barnett, M.K. Sadi Carnot and the Second Law of Thermodynamics. *Osiris* **1958**, *13*, 327–357. [CrossRef]
9. Erlichson, H. Sadi Carnot, 'Founder of the Second Law of Thermodynamics'. *Eur. J. Phys.* **1999**, *20*, 183–192. [CrossRef]
10. Lemons, D.S.; Penner, M.K. Sadi Carnot's contribution to the second law of thermodynamics. *Am. J. Phys.* **2008**, *76*, 21–25. [CrossRef]
11. Feidt, M. The History and Perspectives of Efficiency at Maximum Power of the Carnot Engine. *Entropy* **2017**, *19*, 369. [CrossRef]
12. Huang, K. *Podstawy Fizyki Statystycznej (Inroduction to Statistical Physics)*, 1st ed.; Wydawnictwo Naukowe PWN: Warsaw, Poland, 2006; pp. 16–20.
13. Jakubczyk, P. Toporna logika brudnych maszyn (The primitive logic of dirty machines). *Delta* **2020**, *3*, 1–3.
14. Resnick, R.; Halliday, D. *Fizyka dla StudentóW nauk Przyrodniczych i Technicznych Tom I (Physics for Students of Science and Engineering)*, 2nd ed.; Wydawnictwo Naukowe PWN: Warsaw, Poland, 1966; pp. 697–728.
15. Sawieliew, I.W. *Wykłady z Fizyki, tom 1. Mechanika. Fizyka CząSteczkowa (Lectures on Physics, Volume 1. Mechanics. Molecular Physics)*, 2nd ed.; Wydawnictwo Naukowe PWN: Warsaw, Poland, 1994; pp. 382–394.
16. Bhattacharyya, K. Critique of some thermodynamic proofs based on the pump-engine couple. *arXiv* **2008**, arXiv:0803.3494.
17. Clausius, R. Ueber die bewegende Kraft der Wärme und die Gesetze, welche sich daraus für die Wärmelehre selbst ableiten lassen (On the Moving Force of Heat, and the Laws regarding the Nature of Heat itself which are deducible therefrom). *Ann. Phys.* **1850**, *79*, 368–397, 500–524. [CrossRef]
18. Clausius, R. On the Moving Force of Heat, and the Laws regarding the Nature of Heat itself which are deducible therefrom. *Philos. Mag.* **1851**, *2*, 1–21, 102–119. [CrossRef]
19. Clausius, R. Ueber eine veränderte Form des zweiten Hauptsatzes der mechanischen Wärmetheorie (On a Modified Form of the Second Fundamental Theorem in the Mechanical Theory of Heat). *Ann. Phys.* **1854**, *93*, 481–506. [CrossRef]
20. Clausius, R. On a Modified Form of the Second Fundamental Theorem in the Mechanical Theory of Heat. *Philos. Mag.* **1856**, *12*, 81–98. [CrossRef]
21. Xue, T.-W.; Guo, Z.-Y. What Is the Real Clausius Statement of the Second Law of Thermodynamics? *Entropy* **2019**, *21*, 926. [CrossRef]
22. Clausius, R. *The Mechanical Theory of Heat, with its Applications to the Steam-Engine and to the Physical Properties of Bodies*, 1st ed.; Hirst, T.A., Ed.; John Van Voorst: London, UK, 1867; pp. 1–374.
23. Clausius, R. *The Mechanical Theory of Heat*, 1st ed.; Walter, W.R., Translator; Macmillan & Co.: London, UK, 1879; pp. 1–376.
24. Thomson, W. On the Dynamical Theory of Heat, with numerical results deduced from Mr Joule's Equivalent of a Thermal Unit, and M. Regnault's Observations on Steam (Part I–V). *Trans. R. Soc. Edinb.* **1851**, *20*, 261–298, 475–482. [CrossRef]

25. Thomson, W. On the Dynamical Theory of Heat, with numerical results deduced from Mr Joule's equivalent of a Thermal Unit, and M. Regnault's Observations on Steam (Part I–III). *Philos. Mag. J. Sci.* **1852**, *4*, 8–21, 105–117, 168–256. [CrossRef]
26. Szargut, J. *Termodynamika (Thermodynamics)*, 5th ed.; Wydawnictwo Naukowe PWN: Warsaw, Poland, 1985; pp. 120–122.
27. Pudlik, W. *Termodynamika (Thermodynamics)*, 3rd ed.; Wydawnictwo Politechniki Gdańskiej: Gdańsk, Poland, 2020; pp. 130–187.
28. Kaleta, A.; Górnicki, K. *Podstawy Techniki Cieplnej w InżYnierii Rolniczej (Basics of Thermal Technology in Agricultural Engineering)*, 1st ed.; Wydawnictwo SGGW: Warsaw, Poland, 2009; pp. 105–106.
29. Pippard, A.B.*Elements of Classical Thermodynamics for Advanced Students of Physics*, 3nd ed.; Cambridge University Press: Cambridge, UK, 1966; pp. 29–42.
30. Recknagel, H.; Sprenger, E.; Schramek, E.-R. *Kompendium Ogrzewnictwa i Klimatyzacji (Compendium of Heating and Air Conditioning)*, 1st ed.; OMNI SCALA: Wrocław, Poland, 2008; pp. 103–113.
31. Wiśniewski, S. *Termodynamika techniczna (Technical thermodynamics)*, 5th ed.; Wydawnictwa Naukowo-Techniczne WNT: Warsaw, Poland, 2009; pp. 81–111.
32. Thomson, W. On a Universal Tendency in Nature to the Dissipation of Mechanical Energy. *Philos. Mag. J. Sci.* **1852**, *4*, 304–306. [CrossRef]
33. Radhakrishnamurty, P. Are the second law principles of Caratheodory and Kelvin equivalent? *arXiv* **2011**, arXiv:1102.4235.
34. Hołyst, R.; Poniewierski, A.; Ciach, A. *Termodynamika dla ChemikóW, FizykóW i InżYnieróW (Thermodynamics for Chemists, Physicists and Engineers)*, 1st ed.; Wydawnictwo UKSW: Warsaw, Poland, 2005; pp. 52–62.
35. Clausius, R. Ueber verschiedene für die Anwendung bequeme Formen der Hauptgleichungen der mechanischen Wärmetheorie (About various convenient Forms of the main Equations of Mechanical Heat Theory). *Ann. Phys.* **1865**, *125*, 353–400. [CrossRef]
36. Weber, P.; Bełdowski, P.; Bier, M.; Gadomski, A. Entropy Production Associated with Aggregation into Granules in a Subdiffusive Environment. *Entropy* **2018**, *20*, 651. [CrossRef] [PubMed]
37. Bier, M.; Lisowski, B.; Gudowska-Nowak, E. Phase transitions and entropies for synchronizing oscillators. *Phys. Rev. E* **2016**, *93*, 12143. [CrossRef] [PubMed]
38. Radhakrishnamurty, P. A Critique on Caratheodory Principle of the Second Law of Thermodynamics. *arXiv* **2011**, arXiv:1103.4359.
39. Newelski, L. More on locally atomic models. *Fundam. Math.* **1990**, *136*, 21–26. [CrossRef]
40. Chang, C.C.; Keisler, J. *Model Theory*, 3nd ed.; Dover Publications: New York, NY, USA, 2012; pp. 7–60.
41. Smoluchowski, M. Gültigkeitsgrenzen des zweiten Hauptsatzes des Wärmetheorie (Limits of validity of the second law of thermodynamics). Vorträge über die kinetische Theorie der Materie und der Elektrizität (Lectures on the kinetic theory of matter and electricity). In *Mathematische Vorlesungen an der Universität Göttingen*; B.G. Teubner: Leipzig, Germany, 1914; pp. 87–121.
42. Nikulov, A. The Law of Entropy Increase and the Meissner Effect. *Entropy* **2022**, *24*, 83. [CrossRef]

entropy

Article

Mixedness, Coherence and Entanglement in a Family of Three-Qubit States

Joanna K. Kalaga [1], Wiesław Leoński [1,*], Radosław Szczęśniak [2] and Jan Peřina, Jr. [3]

1 Quantum Optics and Engineering Division, Institute of Physics, University of Zielona Góra, Prof. Z. Szafrana 4a, 65-516 Zielona Góra, Poland; j.kalaga@if.uz.zgora.pl
2 Division of Physics, Częstochowa University of Technology, Ave. Armii Krajowej 19, 42-200 Częstochowa, Poland; radoslaw.szczesniak@pcz.pl
3 Joint Laboratory of Optics of Palacký University and Institute of Physics of CAS, Faculty of Science, Palacký University, 17. listopadu 12, 771 46 Olomouc, Czech Republic; jan.perina.jr@upol.cz
* Correspondence: w.leonski@if.uz.zgora.pl

Abstract: We consider a family of states describing three-qubit systems. We derived formulas showing the relations between linear entropy and measures of coherence such as degree of coherence, first- and second-order correlation functions. We show that qubit–qubit states are strongly entangled when linear entropy reaches some range of values. For such states, we derived the conditions determining boundary values of linear entropy parametrized by measures of coherence.

Keywords: quantum entanglement; linear entropy; coherence; purity of states; concurrence; three-qubit systems

Citation: Kalaga, J.K.; Leoński, W.; Szczęśniak R.; Peřina, J., Jr. Mixedness, Coherence and Entanglement in the Family of Three-Qubit States. *Entropy* **2022**, *24*, 324. https://doi.org/10.3390/e24030324

Academic Editor: Adam Gadomski

Received: 30 December 2021
Accepted: 22 February 2022
Published: 24 February 2022

1. Introduction

Recent developments in modern physics showed that quantum correlations such as quantum entanglement and their relations to quantum coherence play a valid role in understanding the nature of various physical systems.

Coherence is a phenomenon studied not only in classical theories such as ray optics but also is discussed for a variety of quantum systems, for instance, those related to quantum information theory. For the first time, the concept of the degree of coherence was introduced in the area of classical field propagation theory by Zernike in 1938 [1]. Next, in 1950, Hanbury Brown and Twiss investigated the higher-order coherence in the stellar interferometer system [2]. The quantum coherence theory was formulated in 1963 by Glauber [3,4] and Sudarshan [5] and then developed in 1965 by Metha and Sudarshan [6]. On the other hand, we can find an exhaustive presentation of classical and quantum coherence theory in [7] and [8,9], respectively. The quantum coherence theory found numerous applications in research in the field of quantum optics [3,4]. Primarily, in recent years, the relations between quantum coherence and entanglement have been investigated in various models, including those describing atomic ensembles in high-Q cavities [10], optomechanical systems [11], two strongly coupled bosonic modes [12], or three-mode optomechanical systems [13].

The entangled systems found various implementations in the quantum information theory, especially in quantum communication, quantum cryptography [14], and quantum computations [15–22]. The maximally or strongly entangled states play a fundamental role in such processes as quantum teleportation [23–26] or secure quantum communication [27,28]. Thus, it is still essential to deepen knowledge about the nature of entanglement and its relations to other forms of quantum correlations and coherence. Thus, in our research, we will not only consider the relations between entanglement and coherence but also the mixedness of states. The mutual relations between the quantities describing entanglement and mixedness [29–35] or coherence and mixedness [36–41], or coherence and entanglement [42–48] have already been studied in recent years. Our research concerns a three-qubit model that can be implemented in various physical systems. For instance, it could be three two-state spin mutually interacting systems [49] or three two-level atoms [50,51]. In fact, all tripartite systems for which evolution remains closed within a finite set of the states (here, to two states) could be considered in that context. Therefore, our studies are more general, and obtained results can be used in various physical systems.

The paper is organized as follows: in Section 2, we introduce two families of states describing the three-qubit systems of our interest. For such defined groups of states, in Section 3, we study the relations between the mixedness defined by linear entropy and coherence for a qubit–qubit subsystem of our tripartite model. Applying entanglement measures, we find the conditions determining when strongly entangled mixed states appear for the qubit–qubit subsystems. In Sections 4 and 5, for the double excited systems, we analyze the first- and second-order correlation functions, respectively. For two-qubit states, we find possible values of linear entropy parametrized by both correlation functions considered here and derive the formulas which allow identifying ranges of values of discussed parameters for which strongly entangled states can be found.

2. The Three-Qubit System

In this paper, we concentrate on the states describing three-qubit systems (see Figure 1) and studying relations among various quantities describing two-qubit correlations and mixedness of states. The presented analysis is devoted to the bosonic systems that can behave as linear or nonlinear quantum scissors [52]. In other words, the wave function describing the states of such systems is defined in the finite-dimensional Hilbert space [53,54]. Here, we discuss a particular case when only two states are populated for each subsystem. For instance, in the cases of quantum-optical systems, they are vacuum $|0\rangle$ and one-photon $|1\rangle$ states. However, we do not analyze a specific quantum model, but we examine the various states generated in such systems.

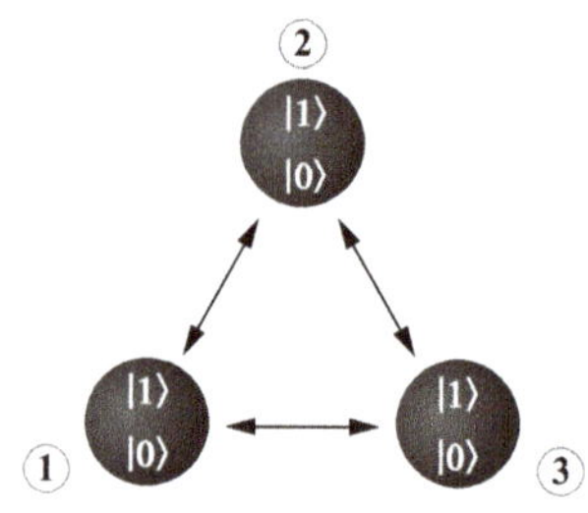

Figure 1. The model of a three-qubit system. The qubits are represented by the black circles and the arrows symbolize the analyzed here the bipartite correlations.

In particular, we shall focus on the two families of states: those corresponding to one excitation in the system and, next, two excitations. First, we concentrate on the situation when we deal with a single excitation, so the total number of photons/phonons $\langle n\rangle = \langle n_1\rangle + \langle n_2\rangle + \langle n_3\rangle = 1$, where indices 1–3 label the qubits. For such a case, the wave function describing the system's state is

$$|\psi\rangle = C_{001}|001\rangle + C_{010}|010\rangle + C_{100}|100\rangle\,, \tag{1}$$

and the corresponding density matrix takes the following form:

$$\rho = |\psi\rangle\langle\psi| = \begin{bmatrix} 0 & 0 & 0 & 0 & 0 & 0 & 0 & 0 \\ 0 & P_{001} & C^*_{001}C_{010} & 0 & C^*_{001}C_{100} & 0 & 0 & 0 \\ 0 & C^*_{010}C_{001} & P_{010} & 0 & C^*_{010}C_{100} & 0 & 0 & 0 \\ 0 & 0 & 0 & 0 & 0 & 0 & 0 & 0 \\ 0 & C^*_{100}C_{001} & C^*_{100}C_{010} & 0 & P_{100} & 0 & 0 & 0 \\ 0 & 0 & 0 & 0 & 0 & 0 & 0 & 0 \\ 0 & 0 & 0 & 0 & 0 & 0 & 0 & 0 \\ 0 & 0 & 0 & 0 & 0 & 0 & 0 & 0 \end{bmatrix}. \tag{2}$$

The C_{ijk} are the complex probability amplitudes corresponding to the states $|ijk\rangle$, whereas $P_{ijk} = C^*_{ijk}C_{ijk}$ are the probabilities related to the latter.

For the second situation that we are interested in, two excitations are present in the system – $\langle n\rangle = \langle n_1\rangle + \langle n_2\rangle + \langle n_3\rangle = 2$. For such a case, we consider the following wave-function:

$$|\psi\rangle = C_{011}|011\rangle + C_{101}|101\rangle + C_{110}|110\rangle, \tag{3}$$

and the corresponding density matrix

$$\rho = |\psi\rangle\langle\psi| = \begin{bmatrix} 0 & 0 & 0 & 0 & 0 & 0 & 0 & 0 \\ 0 & 0 & 0 & 0 & 0 & 0 & 0 & 0 \\ 0 & 0 & 0 & 0 & 0 & 0 & 0 & 0 \\ 0 & 0 & 0 & P_{011} & 0 & C^*_{011}C_{101} & C^*_{011}C_{110} & 0 \\ 0 & 0 & 0 & 0 & 0 & 0 & 0 & 0 \\ 0 & 0 & 0 & C^*_{101}C_{011} & 0 & P_{101} & C^*_{101}C_{110} & 0 \\ 0 & 0 & 0 & C^*_{110}C_{011} & 0 & C^*_{110}C_{101} & P_{110} & 0 \\ 0 & 0 & 0 & 0 & 0 & 0 & 0 & 0 \end{bmatrix}. \tag{4}$$

The two families of states analyzed here are three-qubit states and belong to the same class—that of W-states (for the discussion of various classes of three-qubit states, see [55–57] *and the references quoted therein*). Despite this fact, as we shall show, the values of the first and second-order correlation functions allow for discriminating the states from the two families. Thus, those parameters behave differently from the concurrence and degree of coherence, where those two parameters do not allow for such discrimination. From the other side, the states considered here are those involving one or two excitations. Such states could be physically generated by the systems called *quantum scissors* (both linear and nonlinear ones) [52], and, thus, they seem to be interesting from the practical point of view.

Due to the great attention recently given to W-states [58–64] and a broad range of their application in quantum information systems, we shall focus here on two types of such states. W-states can be employed, for instance, in quantum teleportation systems [65–67], dense coding [68–70], and cryptographic protocols [71,72].

3. The Linear Entropy and Degree of Coherence

In our studies, we concentrate on finding the relation among various quantities characterizing bipartite systems, being subsystems of our three-qubit model. Such two-qubit subsystems appear to be in mixed states. Therefore, one of the quantities analyzed by us is the degree of mixedness. As a measure of mixedness, we will apply the linear entropy defined with the application of purity parameter [31]

$$E(\rho) \equiv \frac{Dim}{Dim-1}\left[1 - Tr\left(\rho^2\right)\right], \tag{5}$$

where Dim denotes the dimension of ρ. In our studies, we analyze the mixedness of two-qubit states. Therefore, we assume that $Dim = 4$ and thus the *linear entropy* can be written as:

$$E_{ij} = E(\rho_{ij}) \equiv \frac{4}{3}\left[1 - Tr\left(\rho_{ij}^2\right)\right]. \tag{6}$$

where ρ_{ij} is the reduced density matrix describing the two-qubit state.

Next, we will analyze the coherence. In this paper, we will study two manifestations of that phenomenon. Firstly, we concentrate on the internal coherence of any two subsystems (from all three), described by the *degree of coherence*. In the next section, we will focus on the mutual coherence—*cross-coherence*.

The degree of coherence that will be applied here can be defined with an application of the degrees of first-order coherence D_i and D_j corresponding to the qubits i and j

$$D_k = \sqrt{2\mathrm{Tr}(\rho_k^2) - 1}, \quad k = i, j = \{1,2,3\}, \tag{7}$$

where ρ_k is the reduced density matrix related to qubit k. Next, the parameter D_k is used to define the *degree of coherence* D_{ij}^2 in the bipartite system [9,73]:

$$D_{ij}^2 = \left(D_i^2 + D_j^2\right)/2. \tag{8}$$

The quantity D_{ij}^2 can be treated as a measure of the total coherence inside the two independently considered subsystems. Thus, D_{ij}^2 is equal to 0 only if both subsystems show no coherence. The states with $D_{ij}^2 = 0$ are the state that gives maximal violation of the CHSH inequality—the Bell states [73].

To find the relations between the values of linear entropy and the degree of coherence for two-qubit mixed states, we have generated 10^6 random three-qubit states defined by the density matrix ρ (2). Next, we have found a reduced density matrix ρ_{ij} representing the two-qubit states discussed by us. Such matrices were derived from the full three-qubit density matrix by tracing out one subsystem—the qubit k. Next, for each qubit–qubit state, we have calculated both linear entropy $E(\rho_{ij})$ and degree of coherence D_{ij}^2. The results showing how the value of linear entropy depends on the values of the degree of coherence for the system involving single excitations are presented in Figure 2. It is interesting that those results are identical to those corresponding to the systems with two excitations and described by the density matrix defined by Equation (4). This is the consequence of the fact that, since the states (2) can be transformed into states (4) by a local unitary transformation, linear entropy and degree of coherence are invariant quantities under a local unitary transformation.

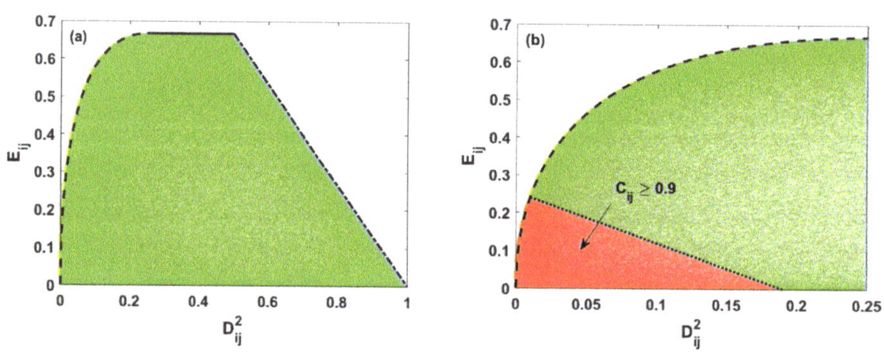

Figure 2. (**a**) Linear entropy E_{ij} versus degree of coherence D_{ij}^2 for two-qubit states described by the density matrix ρ_{ij}, found numerically (green area). Black lines are plotted according to the analytical formulas derived here determining the borders between various regions of the states. (**b**) The same as in (**a**). Additionally, the red area presents the possible values of linear entropy and degree of coherence for two-qubit states with concurrence $C_{ij} > 0.9$ (red area).

For two-qubit mixed states, we see that, for a given value of D_{ij}^2, the linear entropy reaches only some values represented in Figure 2 by the green area. Moreover, the black lines appearing in Figure 2 correspond to the boundary values of E_{ij} defined by Equations (17), (20), and (24).

To find the upper bound of the degree of mixedness for two-qubit states, we express E_{ij} and D_{ij}^2 for each pair of qubits by the probabilities P_{ijk}. For the system described by the density matrix ρ (2), the entropy and degree of coherence are given by (for more details of the calculation method, see in [34,74]):

$$\begin{aligned} E_{12} &\equiv \frac{8}{3}\left(-P_{100}^2 + P_{100} - P_{010}^2 + P_{010} - 2P_{100}P_{010}\right), \\ E_{13} &\equiv \frac{8}{3}\left(-P_{100}^2 + P_{100} - P_{001}^2 + P_{001} - 2P_{100}P_{001}\right), \\ E_{23} &\equiv \frac{8}{3}\left(-P_{010}^2 + P_{010} - P_{001}^2 + P_{001} - 2P_{010}P_{001}\right), \end{aligned} \tag{9}$$

$$\begin{aligned} D_{12}^2 &= 1 + 2\left(P_{100}^2 - P_{100} + P_{010}^2 - P_{010}\right), \\ D_{13}^2 &= 1 + 2\left(P_{100}^2 - P_{100} + P_{001}^2 - P_{001}\right), \\ D_{23}^2 &= 1 + 2\left(P_{010}^2 - P_{010} + P_{001}^2 - P_{001}\right), \end{aligned} \tag{10}$$

whereas, for the double excited system, the formulas describing E_{ij} and D_{ij}^2 take the following forms:

$$\begin{aligned} E_{12} &\equiv \frac{8}{3}\left(-P_{011}^2 + P_{011} - P_{101}^2 + P_{101} - 2P_{011}P_{101}\right), \\ E_{13} &\equiv \frac{8}{3}\left(-P_{011}^2 + P_{011} - P_{110}^2 + P_{110} - 2P_{011}P_{110}\right), \\ E_{23} &\equiv \frac{8}{3}\left(-P_{110}^2 + P_{110} - P_{101}^2 + P_{101} - 2P_{110}P_{101}\right), \end{aligned} \tag{11}$$

$$\begin{aligned} D_{12}^2 &= 1 + 2\left(P_{101}^2 - P_{101} + P_{011}^2 - P_{011}\right), \\ D_{13}^2 &= 1 + 2\left(P_{011}^2 - P_{011} + P_{110}^2 - P_{110}\right), \\ D_{23}^2 &= 1 + 2\left(P_{101}^2 - P_{101} + P_{110}^2 - P_{110}\right). \end{aligned} \tag{12}$$

When $D_{ij}^2 \in \langle 0; 0.25 \rangle$, the maximal values of linear entropy are represented in Figure 2 by the black dashed line. The two-qubit states maximizing the linear entropy for a given value of the degree of coherence are the Werner states. Such states are mixtures of the Bell states and separable ones. The density matrix corresponding to the Werner states discussed here and corresponding to the single excitation's case can be written as:

$$\rho_W = \begin{bmatrix} 1-\alpha & 0 & 0 & 0 \\ 0 & \alpha/2 & \alpha/2 & 0 \\ 0 & \alpha/2 & \alpha/2 & 0 \\ 0 & 0 & 0 & 0 \end{bmatrix}, \tag{13}$$

whereas, for systems with two excitations, has the form:

$$\rho_W = \begin{bmatrix} 0 & 0 & 0 & 0 \\ 0 & \alpha/2 & \alpha/2 & 0 \\ 0 & \alpha/2 & \alpha/2 & 0 \\ 0 & 0 & 0 & 1-\alpha \end{bmatrix}, \tag{14}$$

and the wave-function describing such states is

$$|\psi\rangle = \sqrt{\alpha/2}|\psi_1\rangle + \sqrt{\alpha/2}|\psi_2\rangle + \sqrt{1-\alpha}|\psi_3\rangle, \tag{15}$$

where $\psi_i = \{|001\rangle, |010\rangle, |100\rangle\}$ and $\psi_i = \{|011\rangle, |101\rangle, |110\rangle\}$ for the system with single and double excitation, respectively. The parameter α is related to the probabilities of finding the system in one of these states. Thus, using α, E_{ij} and D^2_{ij} can be expressed as:

$$\begin{aligned} E_{ij} &= \frac{8}{3}\left(\alpha - \alpha^2\right), \\ D^2_{ij} &= 2\left(\frac{\alpha^2}{2} - \alpha\right) + 1. \end{aligned} \tag{16}$$

From Equations (16), we obtain the maximal values of linear entropy for $D^2_{ij} \in \langle 0; 0.25\rangle$ (the black dashed line in Figure 2)

$$E_{ij} = -\frac{8}{3}\left(D^2_{ij} - \sqrt{(D^2_{ij})}\right). \tag{17}$$

In Figure 2, the solid black line represents the maximal value of E_{ij} when $D^2_{ij} \in \langle 0.25; 0.5\rangle$. For such a case, the reduced density matrix ρ_{ij} for the system with a single excitation takes the following form:

$$\rho_{ij} = \begin{bmatrix} 1/2 & 0 & 0 & 0 \\ 0 & \alpha & \sqrt{(1/2-\alpha)\alpha} & 0 \\ 0 & \sqrt{(1/2-\alpha)\alpha} & 1/2-\alpha & 0 \\ 0 & 0 & 0 & 0 \end{bmatrix}, \tag{18}$$

while the density matrix for a double excited system is equal to

$$\rho_{ij} = \begin{bmatrix} 0 & 0 & 0 & 0 \\ 0 & \alpha & \sqrt{(1/2-\alpha)\alpha} & 0 \\ 0 & \sqrt{(1/2-\alpha)\alpha} & 1/2-\alpha & 0 \\ 0 & 0 & 0 & 1/2 \end{bmatrix}, \tag{19}$$

and α reaches values from zero to $1/2$. When $\alpha = 1/4$, the linear entropy $E_{ij} = 2/3$, and the degree of coherence $D^2_{ij} = 1/4$. Whereas, if α is equal to 0 or $1/2$, the linear entropy $E_{ij} = 2/3$ and $D^2_{ij} = 1/2$. For states defined by the density matrix (18) and (19), the linear entropy takes the following form:

$$E_{ij} = \frac{8}{3}\left(\alpha - \alpha^2 - (1/2-\alpha)^2 + 1/2 - \alpha - 2\alpha(1/2-\alpha)\right) = \frac{2}{3}, \tag{20}$$

and does not depend on D^2_{ij}. We note that this value is the maximal value of linear entropy obtained in analyzed families of states.

For the remaining values of degree of coherence D^2_{ij} fulfilling relation $D^2_{ij} > 0.5$, the density matrix ρ_{ij} describing the states corresponding to the maximal values of the linear entropy for single excited states' case is

$$\rho_{ij} = \begin{bmatrix} 1-\alpha-\beta & 0 & 0 & 0 \\ 0 & \alpha & 0 & 0 \\ 0 & 0 & \beta & 0 \\ 0 & 0 & 0 & 0 \end{bmatrix}, \tag{21}$$

while, for the case of the double excitation, it takes the form

$$\rho_{ij} = \begin{bmatrix} 0 & 0 & 0 & 0 \\ 0 & \alpha & 0 & 0 \\ 0 & 0 & \beta & 0 \\ 0 & 0 & 0 & 1-\alpha-\beta \end{bmatrix}. \tag{22}$$

The full density matrix (describing three-qubit system) for such situations is

$$\rho = \alpha|\psi_1\rangle\langle\psi_1| + \beta|\psi_2\rangle\langle\psi_2| + (1-\alpha-\beta)|\psi_3\rangle\langle\psi_3|, \tag{23}$$

where $\psi_i = \{|001\rangle, |010\rangle, |100\rangle\}$ and $\alpha, \beta = \{P_{001}, P_{010}, P_{101}\}$ or $\psi_i = \{|011\rangle, |101\rangle, |110\rangle\}$ and $\alpha, \beta = \{P_{011}, P_{101}, P_{110}\}$ for the system with single and double excitation, respectively, and one of the probabilities, α or β, equals zero. If $\alpha = 0$, the probability β can take values from zero to unity. When β is 0 or 1, the linear entropy reaches zero, and the degree of coherence is equal to 1—while, for $\beta = 1/2$, the linear entropy $E_{ij} = 2/3$ and $D_{ij}^2 = 1/2$.

In fact, the two-qubit states discussed here are the mixtures of two separable states. For such a case, the relation between the linear entropy and the degree of coherence derived for those density matrices using the Formulas (10)–(13) can be expressed as

$$E_{ij} = \frac{4}{3} - \frac{4}{3}D_{ij}^2, \tag{24}$$

which is represented by the dash-dotted line in Figure 2.

In the following steps, we will derive the formula determining the boundary values of linear entropy parametrized by the degree of coherence for the strongly entangled states. In Figure 2b, the red area corresponds to such states, and the dotted line presents such boundary values of linear entropy.

To determine the degree of entanglement between two qubits, we will apply the *concurrence*. The concurrence of the qubit–qubit subsystem can be calculated with the application of the definition proposed by Hill and Wootters [75,76]

$$C_{ij} = C(\rho_{ij}) = max\left(\sqrt{\lambda_I} - \sqrt{\lambda_{II}} - \sqrt{\lambda_{III}} - \sqrt{\lambda_{IV}}, 0\right), \tag{25}$$

where the parameters λ_l are the eigenvalues of matrix R obtained from the relation $R = \rho_{ij}\tilde{\rho_{ij}}$, $\tilde{\rho_{ij}}$ is defined as $\tilde{\rho_{ij}} = \sigma_y \otimes \sigma_y \rho_{ij}^* \sigma_y \otimes \sigma_y$, and σ_y is a 2×2 Pauli matrix.

Next, applying definition (25), we derive the formulas describing concurrence for different pairs of qubits. For the systems with single excitation, concurrence can be expressed by the probabilities as:

$$\begin{aligned} C_{12} &= \sqrt{4P_{100}P_{010}}, \\ C_{13} &= \sqrt{4P_{100}P_{001}}, \\ C_{23} &= \sqrt{4P_{010}P_{001}}, \end{aligned} \tag{26}$$

and, for the double excited system, is

$$\begin{aligned} C_{12} &= \sqrt{4P_{011}P_{101}}, \\ C_{13} &= \sqrt{4P_{011}P_{110}}, \\ C_{23} &= \sqrt{4P_{101}P_{110}}. \end{aligned} \tag{27}$$

In the next step, we shall identify states that are strongly entangled. In our consideration, we assume that the strongly entangled states are those for which the concurrence takes values equal to or higher than 0.9. Applying definition (27,28) and assuming that

$C_{ij} = 0.9$, we can find the relations among probabilities P_{ijk} and obtain the formula that gives the value of the linear entropy represented in Figure 2b by the dotted line:

$$E_{ij} = \frac{19}{75} - \frac{4}{3} D_{ij}^2 . \tag{28}$$

From Figure 2b, we see that the two-qubit states are strongly entangled when the linear entropy and degree of coherence reach small values. More precisely, the strongly entangled states (when $C_{ij} \geq 0.9$) can be generated when the linear entropy becomes equal to or smaller than those defined by Equation (28) for $D_{ij}^2 \in \langle 0.01; 0.19 \rangle$ and when $D_{ij}^2 < 0.01$ by Formula (17).

In three-qubit systems, in addition to entanglement between two qubits, we can also analyze the entanglement of one qubit with the other two. Such entanglement can be quantified by the bipartite concurrence [77]

$$C_{k-ij} = \sqrt{2 - 2Tr(\rho_k^2)} , \tag{29}$$

where ρ_k is the reduced density matrix related to qubit k, and the quantity C_{k-ij} describes entanglement between qubit k and pair of qubits i and j.

The families of states analyzed here are W-class states. For such states, the three-tangle τ_{ijk} that describes the three-way entanglement vanishes. Therefore, using the definition of three-tangle [77],

$$\tau_{ijk} = C_{k-ij}^2 - C_{ik}^2 + C_{jk}^2 , \tag{30}$$

we can write the monogamy relation in the following form:

$$C_{k-ij}^2 = C_{ik}^2 + C_{jk}^2 . \tag{31}$$

The relation (31) can be confirmed using Equations (27), (28) and (29), and is in agreement with the results presented in [77].

Next, applying formulas (10), (12), (27), (28) and (31), we can find the relation between linear entropy E_{ij} and concurrence C_{k-ij}:

$$E_{ij} = \frac{2}{3} C_{k-ij}^2 . \tag{32}$$

Analyzing Equations (27) and (28), we find that maximal value of C_{ik}^2 parametrized by C_{jk}^2 is

$$\max C_{ik}^2 = 1 - C_{jk}^2 , \tag{33}$$

and the maximal reachable value by concurrence C_{k-ij} is 1. Therefore, based on Equation (32), we can confirm that the maximal value of linear entropy obtained in analyzed families of states is 2/3.

4. The First-Order Correlation Function and Linear Entropy

In Section 3, we discussed the relationship between the internal coherence of subsystems (quantified by the degree of coherence D_{ij}^2), linear entropy and concurrence. Here, we shall consider the relationships among the mutual coherence quantified by the first-order correlation function and linear entropy and concurrence. Such first-order cross-correlation function for subsystems i and j can be written as [78,79]:

$$g_{ij}^{(1)} = \frac{|\langle \hat{a}_i^\dagger \hat{a}_j \rangle|}{\sqrt{\langle \hat{a}_i^\dagger \hat{a}_i \rangle \langle \hat{a}_j^\dagger \hat{a}_j \rangle}} . \tag{34}$$

The function $g_{ij}^{(1)}$ can take values from zero to unity. For maximally coherent states, it equals 1, whereas, when we do not observe coherence between subsystems i and j, $g_{ij}^{(1)} = 0$.

All states corresponding to the single excitation's case, described by the wave function (1), are fully coherent and thus $g_{ij}^{(1)} = 1$. In contrast, if we assume the presence of two excitations (see, the wave function (3), the first-order correlation function can take various values from 0 to 1. Therefore, in further analysis, we focus only on the relations between linear entropy and first-order coherence for double excited systems.

In Figure 3, we present the results of numerical analysis concerning the ensemble of randomly generated states describing double excited systems. For such states, the blue area shows possible values of linear entropy for given values of the first-order correlation function. The boundary values of linear entropy are represented by black lines: solid and dashed ones.

To derive the maximal values of linear entropy parametrized by the first-order correlation function, we find the formulas describing $g_{ij}^{(1)}$ function expressed by probabilities:

$$\begin{aligned} g_{12}^{(1)} &= \frac{C_{011}^{*}C_{101}}{\sqrt{(P_{101}+P_{110})(P_{011}+P_{110})}}, \\ g_{13}^{(1)} &= \frac{C_{011}^{*}C_{110}}{\sqrt{(P_{110}+P_{101})(P_{011}+P_{101})}}, \\ g_{23}^{(1)} &= \frac{C_{101}^{*}C_{110}}{\sqrt{(P_{110}+P_{011})(P_{101}+P_{011})}}. \end{aligned} \tag{35}$$

In further analysis, we will consider real probability amplitudes $C_{ijk}^{*} = C_{ijk} = \sqrt{P_{ijk}}$.

From Figure 3, we see that, for $g_{ij}^{(1)} \leq 1/3$, the maximal value of E_{ij} does not depend on the value of the first-order correlation function. For such a case, the two-qubit matrix is expressed by Equation (19), and the corresponding first-order correlation function is given as

$$g_{ij}^{(1)} = \frac{\sqrt{(1/2-\alpha)\alpha}}{\sqrt{(1-\alpha)(\alpha+1/2)}}. \tag{36}$$

Thus, for $g_{ij}^{(1)} \leq 1/3$, the maximal value of E_{ij} is equal to $2/3$ and does not depend on the values of the parameter α.

From the other side, when $g_{ij}^{(1)} > 1/3$, the maximal possible value of linear entropy decreases with the increasing value of the first-order correlation function (see the dashed line in Figure 3). In such a case, the density matrix describing the system is:

$$\rho_{ij} = \begin{bmatrix} 0 & 0 & 0 & 0 \\ 0 & \alpha & \sqrt{\alpha\beta} & 0 \\ 0 & \sqrt{\alpha\beta} & \beta & 0 \\ 0 & 0 & 0 & 1-\alpha-\beta \end{bmatrix}, \tag{37}$$

where the probabilities α and β have to be equal to

$$\alpha = \beta = \frac{g_{ij}^{(1)}}{1+g_{ij}^{(1)}}, \tag{38}$$

and the probabilities α and β can take values within the range $\langle 1/4, 1/2 \rangle$. When $\alpha = \beta = 1/4$, the first-order correlation function is $1/3$, and $E_{ij} = 2/3$. However, if $\alpha = \beta = 1/2$, the linear entropy reaches zero, and function $g_{ij}^{(1)}$ is equal to unity.

In general, for the two-qubit states represented by Equation (37), E_{ij} fulfills the following relation:

$$E_{ij} = -\frac{16\left(g_{ij}^{(1)} - 1\right)g_{ij}^{(1)}}{3\left(1 + g_{ij}^{(1)}\right)^2}. \quad (39)$$

In the next step, we discuss the case when the states are strongly entangled ones, i.e., the concurrence is assumed to be equal to or higher than 0.9. For such a situation, the minimal value of E_{ij} parametrized by $g_{ij}^{(1)}$ is defined by the condition represented by the dash-dotted line in Figure 3b. The red area corresponds to the values linear entropy and first-order correlation function for states presenting strong entanglement. From Figure 3b, we see that the states with $C_{ij} \geq 0.9$ exhibit a high level of the first-order correlation function $g_{ij}^{(1)} \in \langle 9/11; 1\rangle$. Moreover, for such the case, the linear entropy is limited to values determined by:

$$E_{ij} \geq -\frac{27\left(g_{ij}^{(1)^2} - 1\right)\left(481 g_{ij}^{(1)^2} - 81\right)}{20000 g_{ij}^{(1)^4}}. \quad (40)$$

We derived that condition using the definitions (12), (28) and (36) and assuming that $C_{ij} = 0.9$.

Thus, one can state that the strongly entangled two-qubit states are simultaneously characterized by low levels of mixedness and high values of the first-order coherence function.

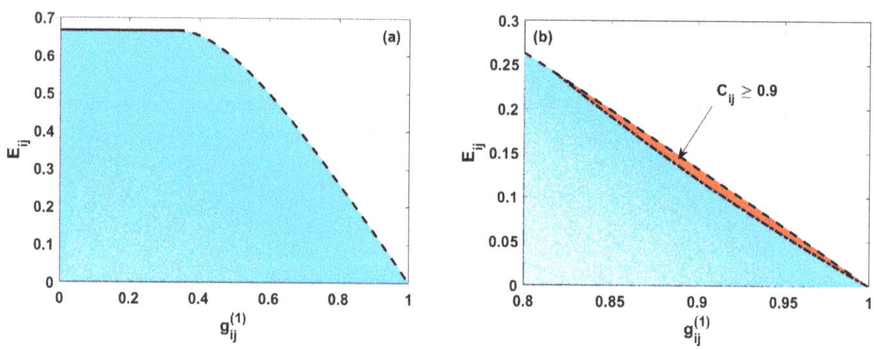

Figure 3. (**a**) Linear entropy E_{ij} versus first-order correlation function $g_{ij}^{(1)}$ for two-qubit states described by the density matrix ρ_{ij}, calculated numerically (blue area). Black lines are plotted according to the analytical formulas derived here determining the borders between various regions of the states. (**b**) The same as in (**a**). Additionally, the red area presents the possible values of linear entropy and the first-order correlation function for two-qubit states with concurrence $C_{ij} > 0.9$ (red area).

5. The Second-Order Correlation Function and Linear Entropy

Analogously, as in the previous section, we will analyze at this point relations between the degree of mixedness and second-order coherence function $g_{ij}^{(2)}$. This function quantifies the correlations between intensities of field, contrary to $g_{ij}^{(1)}$ considered earlier that described the correlations between the amplitudes of two fields. $g_{ij}^{(2)}$ is defined here for two subsystems i and j and can be expressed as [78,79]:

$$g_{ij}^{(2)} = \frac{\langle \hat{a}_i^\dagger \hat{a}_j^\dagger \hat{a}_i \hat{a}_j \rangle}{\langle \hat{a}_i^\dagger \hat{a}_i \rangle \langle \hat{a}_j^\dagger \hat{a}_j \rangle}. \quad (41)$$

Applying the procedure described in the previous section, we shall concentrate here on the case of double excited systems described by the density matrix (4). For such a situation, the second-order correlation function expressed by probabilities for each qubit–qubit subsystem can be written as:

$$\begin{aligned} g_{12}^{(2)} &= \frac{P_{110}}{(P_{101}+P_{110})(P_{011}+P_{110})}, \\ g_{13}^{(2)} &= \frac{P_{011}}{(P_{110}+P_{101})(P_{011}+P_{101})}, \\ g_{23}^{(2)} &= \frac{P_{101}}{(P_{110}+P_{011})(P_{101}+P_{011})}. \end{aligned} \tag{42}$$

Figure 4 depicts numerical results of analysis of randomly generated states for the system with double excitation. The same as previously, colored areas correspond to the possible achievable states characterized by various pairs of the values of the linear entropy and $g_{ij}^{(2)}$. The black lines appearing there denote the boundary values of the entropy for the particular $g_{ij}^{(2)}$. When $g_{ij}^{(2)} < 8/9$, the maximal possible value of E_{ij} monotonously increases with the increasing value of the second-order correlation function (see the dashed line in Figure 4). In such a case, using Equations (12) and (43), we find that the maximal value of E_{ij} fulfills the relation:

$$E_{ij} = \frac{16\left(\sqrt{1-g_{ij}^{(2)}}-1\right)^2\left(\sqrt{1-g_{ij}^{(2)}}+g_{ij}^{(2)}-1\right)}{3g_{ij}^{(2)2}}. \tag{43}$$

The entropy E_{ij} given by (43) reaches its maximal values when the system is described by the density matrix (37) with the probabilities α and β equal to:

$$\alpha = \beta = \frac{g_{ij}^{(2)}+\sqrt{1-g_{ij}^{(2)}}-1}{g_{ij}^{(2)}} \tag{44}$$

where α and β can take values in the range $\langle 0;1/2\rangle$. When both α and β are simultaneously equal to 0 or 1/2, the second-order correlation function and entropy become equal to zero. However, if $\alpha = \beta = 1/4$, the linear entropy $E_{ij} = 2/3$, and $g_{ij}^{(2)}$ reaches $= 8/9$.

However, when $g_{ij}^{(2)} \geq 8/9$, the maximal possible value of linear entropy stops being dependent on the second-order correlation function and remains equal to 2/3 (see the black solid line in Figure 4. For such a case, the two-qubit density matrix is described by Equation (19).

In Figure 4b, the red area corresponds to the strongly entangled states with concurrence $C_{ij} \geq 0.9$. The dash-dotted line appearing there represents the condition for the minimal values of E_{ij} parametrized by $g_{ij}^{(2)}$. Simultaneous analysis of Equations (12), (28) and (43), describing the entropy, second-order correlation function, concurrence, respectively, and assuming that concurrence is equal to 0.9 gives us the minimal achievable entropy for strongly entangled states:

$$E_{ij} = \frac{27\left(400-481g_{ij}^{(2)}\right)g_{ij}^{(2)}}{20000\left(g_{ij}^{(2)}-1\right)^2}, \tag{45}$$

where $g_{ij}^{(2)} \in \langle 0; 40/121 \rangle$. It is seen that the strongly entangled two-qubit states are characterized by simultaneously low levels of both mixedness and second-order coherence function.

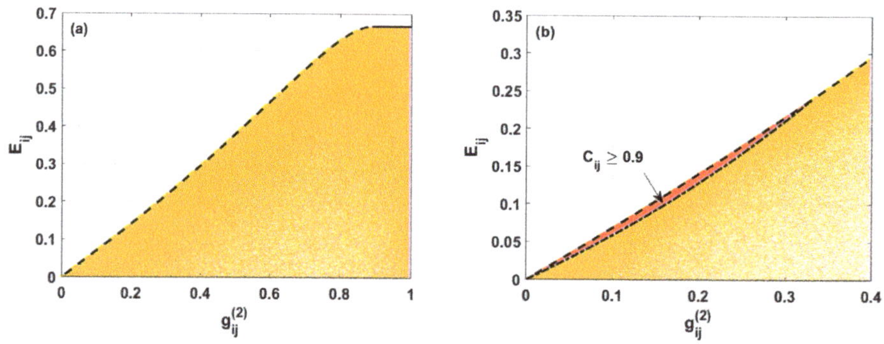

Figure 4. (**a**) Linear entropy E_{ij} versus second-order correlation function $g_{ij}^{(2)}$ for two-qubit states described by the density matrix ρ_{ij}, calculated numerically (yellow area). Black lines are plotted according to the analytical formulas derived here and determining the borders between various regions of the states. (**b**) The same as in (**a**). Additionally, the red area presents the possible values of linear entropy and the second-order correlation function for two-qubit states with concurrence $C_{ij} > 0.9$ (red area).

6. Conclusions

In this work, we have analyzed two families of three-qubit states in the context of the appearance of coherence and entanglement as quantum resources, and the mixedness of discussed states. In particular, we have focused on the characteristics of possible achievable states describing the two-qubit subspace of the system. Applying the tracing out procedure, we have analyzed the degree of mixedness of such two-qubit states, the bipartite coherences, and entanglement. We have compared the degree of mixedness and the parameters describing coherences, such as the degree of coherence, the first- and second-order correlation function, and have shown the relations among them. Based on such performed analysis, we have derived boundary conditions for possible achievable strongly entangled two-qubit states. We have shown that the strongly entangled states can be characterized by low levels of mixedness and degree of coherence. On the other hand, analyzing the correlation functions $g_{ij}^{(1)}$ and $g_{ij}^{(2)}$, it turned out that highly entangled states are states with high and low levels of the first and second-order correlation function, respectively.

Author Contributions: Conceptualization, J.K.K. and W.L.; methodology, J.K.K. and J.P.J.; software, J.K.K.; validation, J.K.K. and R.S.; formal analysis, J.K.K. and W.L.; investigation, J.K.K., W.L., R.S. and J.P.J.; writing—original draft preparation, J.K.K. and W.L.; writing—review and editing, J.K.K., R.S. and J.P.J. All authors have read and agreed to the published version of the manuscript.

Funding: J.K.K. and W.L. acknowledge the support of the program of the Polish Minister of Science and Higher Education under the name "Regional Initiative of Excellence" in 2019-2022, Project No. 003/RID/2018/19, funding amount 11 936 596.10 PLN.

Institutional Review Board Statement: Not applicable.

Informed Consent Statement: Not applicable.

Data Availability Statement: Not applicable.

Conflicts of Interest: The authors declare no conflict of interest.

References

1. Zernike, F. The concept of degree of coherence and its application to optical problems. *Physica* **1938**, *5*, 785–795. [CrossRef]
2. Hanbury Brown, R.; Twiss, R.Q. A Test of a New Type of Stellar Interferometer on Sirius. *Nature* **1956**, *178*, 1046–1048. [CrossRef]
3. Glauber, R.J. The Quantum Theory of Optical Coherence. *Phys. Rev.* **1963**, *130*, 2529–2539. [CrossRef]
4. Glauber, R.J. Coherent and Incoherent States of the Radiation Field. *Phys. Rev. A* **1963**, *131*, 2766–2788. [CrossRef]
5. Sudarshan, E.C.G. Equivalence of Semiclassical and Quantum Mechanical Descriptions of Statistical Light Beams. *Phys. Rev. Lett.* **1963**, *10*, 277–279. [CrossRef]
6. Mehta, C.L.; Sudarshan, E.C.G. Relation between Quantum and Semiclassical Description of Optical Coherence. *Phys. Rev.* **1965**, *138*, B274–B280. [CrossRef]
7. Peřina, J. *Coherence of Light*; Kluwer: Dordrecht, The Netherlands, 1985.
8. Peřina, J. *Quantum Statistics of Linear and Nonlinear Optical Phenomena*; Kluwer: Dordrecht, The Netherlands, 1991.
9. Mandel, L.; Wolf, E. *Optical Coherence and Quantum Optics*; Cambridge University Press: Cambridge, UK, 1995.
10. Sun, L.; Li, G.; Gu, W.; Ficek, Z. Generating coherence and entanglement with a finite-size atomic ensemble in a ring cavity. *New J. Phys.* **2011**, *13*, 093019. [CrossRef]
11. Sun, L.; Li, G.; Ficek, Z. First-order coherence versus entanglement in a nanomechanical cavity. *Phys. Rev. A* **2012**, *85*, 022327. [CrossRef]
12. Bougouffa, S.; Ficek, Z. Evidence of indistinguishability and entanglement determined by the energy-time uncertainty principle in a system of two strongly coupled bosonic modes. *Phys. Rev. A* **2016**, *93*, 063848. [CrossRef]
13. Liu, J.; Zhou, Y.H.; Huang, J.; Huang, J.F.; Liao, J.Q. Quantum simulation of a three-mode optomechanical system based on the Fredkin-type interaction. *Phys. Rev. A* **2021**, *104*, 053715. [CrossRef]
14. Gisin, N.; Ribordy, G.; Tittel, W.; Zbinden, H. Quantum cryptography. *Rev. Mod. Phys.* **2002**, *74*, 145–195. [CrossRef]
15. Ekert, A.K. Quantum cryptography based on Bell's theorem. *Phys. Rev. Lett.* **1991**, *67*, 661–663. doi: [CrossRef] [PubMed]
16. Bennett, C.H.; Wiesner, S.J. Communication via one- and two-particle operators on Einstein-Podolsky-Rosen states. *Phys. Rev. Lett.* **1992**, *69*, 2881–2884. [CrossRef] [PubMed]
17. Bennett, C.H.; Brassard, G.; Crépeau, C.; Jozsa, R.; Peres, A.; Wootters, W.K. Teleporting an unknown quantum state via dual classical and Einstein-Podolsky-Rosen channels. *Phys. Rev. Lett.* **1993**, *70*, 1895–1899. [CrossRef]
18. Grover, L.K. Quantum Mechanics Helps in Searching for a Needle in a Haystack. *Phys. Rev. Lett.* **1997**, *79*, 325–328. [CrossRef]
19. Zheng, S.B.; Guo, G.C. Efficient Scheme for Two-Atom Entanglement and Quantum Information Processing in Cavity QED. *Phys. Rev. Lett.* **2000**, *85*, 2392–2395. [CrossRef] [PubMed]
20. Nielsen, M.A.; Chuang, I.L. *Quantum Computation and Quantum Information*, 10th ed.; Cambridge University Press: Cambridge, UK, 2010.
21. Rieffel, E.G.; Polak, W.H. *Quantum Computing: A Gentle Introduction*; MIT Press: Cambridge, MA, USA, 2011.
22. Estarellas, M.P.; D'Amico, I.; Spiller, T.P. Robust quantum entanglement generation and generation-plus-storage protocols with spin chains. *Phys. Rev. A* **2017**, *95*, 042335. [CrossRef]
23. Bennett, C.H.; Brassard, G.; S.Popescu.; Schumacher, B.; Wootters, W.K. Purification of noisy entanglement and faithful teleportation via noisy channels. *Phys. Rev. Lett.* **1996**, *76*, 722. [CrossRef]
24. Boschi, D.; Branca, S.; De Martini, F.; Hardy, L.; Popescu, S. Experimental Realization of Teleporting an Unknown Pure Quantum State via Dual Classical and Einstein-Podolsky-Rosen Channels. *Phys. Rev. Lett.* **1998**, *80*, 1121–1125. [CrossRef]
25. Bouwmeester, D.; Pan, J.W.; Mattle, K.; Eible, M.; Weinfurter, H.; Zeilinger, A. Experimental quantum teleportation. *Nature* **1997**, *390*, 575. [CrossRef]
26. Özdemir, Ş.K.; Bartkiewicz, K.; Liu, Y.X.; Miranowicz, A. Teleportation of qubit states through dissipative channels: Conditions for surpassing the no-cloning limit. *Phys. Rev. A* **2007**, *76*, 042325. [CrossRef]
27. He, Q.; Rosales-Zárate, L.; Adesso, G.; Reid, M.D. Secure Continuous Variable Teleportation and Einstein-Podolsky-Rosen Steering. *Phys. Rev. Lett.* **2015**, *115*, 180502. [CrossRef] [PubMed]
28. Kocsis, S.; Hall, M.J.W.; Bennet, A.J.; Saunders, D.J.; Pryde, G.J. Experimental measurement-device-independent verification of quantum steering. *Nat. Commun.* **2015**, *6*, 5886. [CrossRef]
29. Ishizaka, S.; Hiroshima, T. Maximally entangled mixed states under nonlocal unitary operations in two qubits. *Phys. Rev. A* **2000**, *62*, 022310. [CrossRef]
30. Munro, W.J.; James, D.F.V.; White, A.G.; Kwiat, P.G. Maximizing the entanglement of two mixed qubits. *Phys. Rev. A* **2001**, *64*, 030302. [CrossRef]
31. Wei, T.C.; Nemoto, K.; Goldbart, P.M.; Kwiat, P.G.; Munro, W.J.; Verstraete, F. Maximal entanglement versus entropy for mixed quantum states. *Phys. Rev. A* **2003**, *67*, 022110. [CrossRef]
32. Bandyopadhyay, S. Entanglement, mixedness, and perfect local discrimination of orthogonal quantum states. *Phys. Rev. A* **2012**, *85*, 042319. [CrossRef]
33. Paulson, K.; Satyanarayana, S. Bounds on mixedness and entanglement of quantum teleportation resources. *Phys. Lett. A* **2017**, *381*, 1134–1137. [CrossRef]
34. Kalaga, J.K.; Leoński, W. Quantum steering borders in three-qubit systems. *Quantum Inf. Process.* **2017**, *16*, 175. [CrossRef]

35. Ashrafpour, M.; Saghi, Z. Research Paper: Investigation of the Relationship of the Degree of Mixedness and Entanglement of the Bipartite Spin Coherent States. *Iran. J. Appl. Phys.* **2021**, *11*, 7–15. [CrossRef]
36. Cheng, S.; Hall, M.J.W. Complementarity relations for quantum coherence. *Phys. Rev. A* **2015**, *92*, 042101. [CrossRef]
37. Singh, U.; Bera, M.N.; Dhar, H.S.; Pati, A.K. Maximally coherent mixed states: Complementarity between maximal coherence and mixedness. *Phys. Rev. A* **2015**, *91*, 052115. [CrossRef]
38. Rastegin, A.E. Quantum-coherence quantifiers based on the Tsallis relative α entropies. *Phys. Rev. A* **2016**, *93*, 032136. [CrossRef]
39. Yao, Y.; Dong, G.H.; Xiao, X.; Sun, C.P. Frobenius-norm-based measures of quantum coherence and asymmetry. *Sci. Rep.* **2016**, *6*, 32010. [CrossRef]
40. Dixit, K.; Naikoo, J.; Banerjee, S.; Alok, A.K. Study of coherence and mixedness in meson and neutrino systems. *Eur. Phys. J. C* **2019**, *79*, 96. [CrossRef]
41. Radhakrishnan, C.; Ding, Z.; Shi, F.; Du, J.; Byrnes, T. Basis-independent quantum coherence and its distribution. *Ann. Phys.* **2019**, *409*, 167906. [CrossRef]
42. Streltsov, A.; Singh, U.; Dhar, H.S.; Bera, M.N.; Adesso, G. Measuring Quantum Coherence with Entanglement. *Phys. Rev. Lett.* **2015**, *115*, 020403. [CrossRef]
43. Xi, Z.; Li, Y.; Fan, H. Quantum coherence and correlations in quantum system. *Sci. Rep.* **2015**, *5*, 10922. [CrossRef]
44. Ma, J.; Yadin, B.; Girolami, D.; Vedral, V.; Gu, M. Converting Coherence to Quantum Correlations. *Phys. Rev. Lett.* **2016**, *116*, 160407. [CrossRef]
45. Killoran, N.; Steinhoff, F.E.S.; Plenio, M.B. Converting Nonclassicality into Entanglement. *Phys. Rev. Lett.* **2016**, *116*, 080402. doi: [CrossRef]
46. Chitambar, E.; Hsieh, M.H. Relating the Resource Theories of Entanglement and Quantum Coherence. *Phys. Rev. Lett.* **2016**, *117*, 020402. [CrossRef] [PubMed]
47. Goswami, S.; Adhikari, S.; Majumdar, A.S. Coherence and entanglement under three-qubit cloning operations. *Quantum Inf. Process.* **2018**, *18*, 36. [CrossRef]
48. Shamsi, Z.H.; Noreen, A.; Mushtaq, A. Analysis of quantum coherence for localized fermionic systems in an accelerated motion. *Results Phys.* **2020**, *19*, 103302. [CrossRef]
49. Fei, J.; Zhou, D.; Shim, Y.; Oh, S.; Hu, X.; Friesen, M. Mediated gates between spin qubits. *Phys. Rev. A* **2012**, *86*, 062328. [CrossRef]
50. Iqbal, M.S.; Mahmood, S.; Razmi, M.S.K.; Zubairy, M.S. Interaction of two two-level atoms with a single-mode quantized radiation field. *J. Opt. Soc. Am. B* **1988**, *5*, 1312–1316. [CrossRef]
51. Ficek, Z.; Tanaś, R. Entangled states and collective nonclassical effects in two-atom systems. *Phys. Rep.* **2002**, *372*, 369–443. [CrossRef]
52. Leoński, W.; Kowalewska-Kudłaszyk, A. Quantum Scissors—Finite-Dimensional States Engineering. In *Progress in Optics*; Wolf, E., Ed.; Elsevier: Amsterdam, The Netherlands, 2011; Volume 56, pp. 131–185. [CrossRef]
53. Leoński, W.; Tanaś, R. Possibility of producing the one-photon state in a kicked cavity with a nonlinear Kerr medium. *Phys. Rev. A* **1994**, *49*, R20–R23. [CrossRef]
54. Leoński, W.; Miranowicz, A. Kerr nonlinear coupler and entanglement. *J. Opt. B* **2004**, *6*, S37–S42. [CrossRef]
55. Acín, A.; Andrianov, A.; Jané, E.; Tarrach, R. Three-qubit pure-state canonical forms. *J. Phys. A Math. Gen.* **2001**, *34*, 6725–6739. doi: [CrossRef]
56. Sabin, C.; Garcia-Alcaine, G. A classification of entanglement in three-qubit systems. *Eur. Phys. J. D* **2008**, *48*, 435–442. [CrossRef]
57. Enríquez, M.; Delgado, F.; Życzkowski, K. Entanglement of Three-Qubit Random Pure States. *Entropy* **2018**, *20*, 745. [CrossRef]
58. Cao, Z.L.; Song, W. Teleportation of a two-particle entangled state via W class states. *Phys. A Stat. Mech. Its Appl.* **2005**, *347*, 177–183. [CrossRef]
59. Li, L.; Qiu, D. The states of W-class as shared resources for perfect teleportation and superdense coding. *J. Phys. A Math. Theor.* **2007**, *40*, 10871–10885. [CrossRef]
60. Kim, J.S.; Sanders, B.C. Generalized W-class state and its monogamy relation. *J. Phys. A Math. Theor.* **2008**, *41*, 495301. doi: [CrossRef]
61. Adhikari, S.; Gangopadhyay, S. A Study of the Efficiency of the Class of W-States as a Quantum Channel. *Int. J. Theor. Phys.* **2009**, *48*, 403–408. [CrossRef]
62. Cui, W.; Chitambar, E.; Lo, H.K. Randomly distilling W-class states into general configurations of two-party entanglement. *Phys. Rev. A* **2011**, *84*, 052301. [CrossRef]
63. Liang, Y.; Zheng, Z.J.; Zhu, C.J. Monogamy and polygamy for generalized W-class states using Rényi-α entropy. *Phys. Rev. A* **2020**, *102*, 062428. [CrossRef]
64. Shi, X.; Chen, L. Monogamy relations for the generalized W-class states beyond qubits. *Phys. Rev. A* **2020**, *101*, 032344. doi: [CrossRef]
65. Shi, B.S.; Tomita, A. Teleportation of an unknown state by W state. *Phys. Lett. A* **2002**, *296*, 161–164. [CrossRef]
66. Joo, J.; Park, Y.J. Comment on "Teleportation of an unknown state by W states": [Phys. Lett. A 296 (2002) 161]. *Phys. Lett. A* **2002**, *300*, 324–326. [CrossRef]
67. Shi, B.S.; Tomita, A. Reply to "Comment on: Teleportation of an unknown state by W state": [Phys. Lett. A 300 (2002) 324]. *Phys. Lett. A* **2002**, *300*, 538–539. [CrossRef]
68. Yang, X.; q. Bai, M.; w. Mo, Z. Controlled Dense Coding with the W State. *Int. J. Theor. Phys.* **2017**, *56*, 3525–3533. [CrossRef]

69. Zhou, Y.S.; Wang, F.; Luo, M.X. Efficient Superdense Coding with W States. *Int. J. Theor. Phys.* **2018**, *57*, 1935–1941. [CrossRef]
70. Roy, S.; Chanda, T.; Das, T.; Sen(De) , A.; Sen, U. Deterministic quantum dense coding networks. *Phys. Lett. A* **2018**, *382*, 1709–1715. [CrossRef]
71. Jian, W.; Quan, Z.; Chao-Jing, T. Quantum Secure Communication Scheme with W State. *Commun. Theor. Phys.* **2007**, *48*, 637–640. doi: [CrossRef]
72. Chen, X.B.; Wen, Q.Y.; Guo, F.Z.; Sun, Y.; Xu, G.; Zhu, F.C. Controlled quantum secure direct communication with W state. *Int. J. Quantum Inf.* **2008**, *6*, 899–906. [CrossRef]
73. Svozilík, J.; Vallés, A.; Peřina, J.; Torres, J.P. Revealing Hidden Coherence in Partially Coherent Light. *Phys. Rev. Lett.* **2015**, *115*, 220501. [CrossRef]
74. Kalaga, J.K.; Leoński, W.; Peřina, J. Einstein-Podolsky-Rosen steering and coherence in the family of entangled three-qubit states. *Phys. Rev. A* **2018**, *97*, 042110. [CrossRef]
75. Hill, S.; Wootters, W.K. Entanglement of a Pair of Quantum Bits. *Phys. Rev. Lett.* **1997**, *78*, 5022–5025. [CrossRef]
76. Wootters, W.K. Entanglement of Formation of an Arbitrary State of Two Qubits. *Phys. Rev. Lett.* **1998**, *80*, 2245–2248. [CrossRef]
77. Coffman, V.; Kundu, J.; Wootters, W.K. Distributed entanglement. *Phys. Rev. A* **2000**, *61*, 052306. [CrossRef]
78. Gerry, C.C.; Knight, P.L. *Introductory Quantum Optics*; Cambridge University Press: Cambridge, UK, 2005.
79. Walls, D.F.; Milburn, G.J. *Quantum Optics*, 2nd ed.; Springer: Berlin/Heidelberg, Gemany, 2008.

entropy

Article

Coherence and Anticoherence Induced by Thermal Fields

Lihui Sun [1], Ya Liu [1], Chen Li [1], Kaikai Zhang [1], Wenxing Yang [1,2,*] and Zbigniew Ficek [3,*]

1 Institute of Quantum Optics and Information Photonics, School of Physics and Optoelectronic Engineering, Yangtze University, Jingzhou 434023, China; lhsun@yangtzeu.edu.cn (L.S.); 2021710172@yangtzeu.edu.cn (Y.L.); 202006957@yangtzeu.edu.cn (C.L.); 202073044@yangtzeu.edu.cn (K.Z.)
2 Department of Physics, Southeast University, Nanjing 211189, China
3 Quantum Optics and Engineering Division, Institute of Physics, University of Zielona Góra, Szafrana 4a, 65-516 Zielona Góra, Poland
* Correspondence: wenxingyang@seu.edu.cn (W.Y.); ficekkacst@gmail.com (Z.F.)

Abstract: Interesting coherence and correlations appear between superpositions of two bosonic modes when the modes are parametrically coupled to a third intermediate mode and are also coupled to external modes which are in thermal states of unequal mean photon numbers. Under such conditions, it is found that one of linear superpositions of the modes, which is effectively decoupled from the other modes, can be perfectly coherent with the other orthogonal superposition of the modes and can simultaneously exhibit anticoherence with the intermediate mode, which can give rise to entanglement between the modes. It is shown that the coherence effects have a substantial effect on the population distribution between the modes, which may result in lowering the population of the intermediate mode. This shows that the system can be employed to cool modes to lower temperatures. Furthermore, for appropriate thermal photon numbers and coupling strengths between the modes, it is found that entanglement between the directly coupled superposition and the intermediate modes may occur in a less restricted range of the number of the thermal photons such that the modes could be strongly entangled, even at large numbers of the thermal photons.

Keywords: coherence; anticoherence; entanglement; nonlinear systems

Citation: Sun, L.; Liu, Y.; Li, C.; Zhang, K.; Yang, W.; Ficek, Z. Coherence and Anticoherence Induced by Thermal Fields. *Entropy* **2022**, *24*, 692. https://doi.org/10.3390/e24050692

Academic Editor: Adam Gadomski

Received: 25 April 2022
Accepted: 12 May 2022
Published: 13 May 2022

1. Introduction

The problem of the creation of coherence and correlations between quantum systems has attracted considerable interest over the years not only because of a basic desire to understand how coherence and correlations could be created but also because of their importance in determination of nonclassical states of quantum systems [1–4]. Various types of correlations can exist between quantum systems, and their importance in understanding properties of quantum systems is often discussed in connection with different phenomena. For example, interference and quantum beats are among the simplest examples of phenomena resulting from the presence of mutual coherence, the so-called first-order correlation between quantum systems. Nonclassical phenomena, such as squeezing and entanglement, result from the presence of a different kind of correlation, often referred to as anomalous correlations [5,6]. The mutual coherence resulting from the first-order correlation is produced by a constant or nearly constant phase difference between quantum systems [3,4,7,8]. There are, however, coherence effects resulting from higher-order correlations, e.g., the intensity correlations, which are possible even when the phase difference between systems is random [9–12].

Anomalous and intensity correlations are the natural products of a range of two-photon processes in which simultaneous or nearly simultaneous pairs of photons are produced [13,14]. Because each photon in the pair has no definite phase, there is no constant phase relation between them. Therefore, photons in the pair behave as mutually incoherent. This property has been observed experimentally in the process of parametric down conversion where pairs of photons, called the signal and idler photons, are produced [15,16]. Although the signal and idler photons are mutually incoherent, they are found in an entangled state which results from the anomalous correlation between the photons [17,18]. This observation suggests that the first-order correlation, which is responsible for the coherence and the anomalous correlation, are mutually exclusive. Following this observation, Mandel [19] proposed to call quantum systems exhibiting anomalous correlation as anticoherent.

The purpose of the present paper is to explore further possibilities to create coherence and anticoherence in a multipartite system. We consider a tripartite system composed of three coupled bosonic modes and investigate their coherence and anicoherence properties in an example of a three-mode optomechanical system, which consists of two cavity modes simultaneously coupled to the mode of a vibrating membrane. We assume that the cavity modes are affected by external input modes, which are in thermal states of unequal mean photon numbers. The difference in the mean number of photons of the input thermal fields constitutes an important and essential aspect of the work presented here. We will show how the populations of the modes and the correlations between them are sensitive to the population of the external thermal modes. When the external modes are in thermal states of different mean number of photons, we find that the steady-state populations of the modes can be dramatically altered, even to the point of the complete transfer of the population between the modes. Moreover, coherence and anticoherence, which may lead to entanglement between modes, can be established between modes which are completely decoupled from each other. This is certainly a surprising result since one would expect no correlations between decoupled modes affected by external thermal fields.

The paper is organized as follows. In Section 2, we introduce our model and the method of the evaluation of the dynamics of the systemś modes using an optomechanical system as an illustration. In Section 3, we study the properties of the steady-state population distribution between the modes. Section 4 is devoted to studying the correlations between the modes. We finish in Section 5 with the conclusion. In Appendix A, we present, as an illustration, a detailed derivation of the analytical expression for the steady-state population of the membrane mode.

2. Three-Mode System

The system we study consists of three parts; two modes whose fields are described by annihilation operators a_1 and a_2, coupled to a third mode whose field is described by an annihilation operator b. The modes a_1 and a_2 are coupled to mode b through the nonlinear (parametric) squeezing-type interactions. There is no direct coupling between modes a_1 and a_2. The Hamiltonian interaction for the three coupled modes is taken to be

$$H = \hbar g_1 \left(a_1^\dagger b^\dagger + a_1 b \right) + \hbar g_2 \left(a_2^\dagger b^\dagger + a_2 b \right), \tag{1}$$

where g_1 and g_2 are the coupling constants between modes a_1 and b, and a_2 and b, respectively. The nonlinear squeezing-type interactions, as described by the Hamiltonian (1) can be created in a variety of systems. For example, squeezing-type interactions between several modes have been realized in linear optical schemes involving external source of squeezed light and networks of beamsplitters [20]. Another example where this type of interaction can be created is a ring cavity containing an atomic ensemble coupled to counter-propagating modes of the cavity [21,22].

A good example of such a system is an optomechanical system consisting of two single-mode cavities sharing an oscillating mirror [23], or a single-mode optical cavity coupled to two mechanical modes of a vibrating membrane [24–26]. The method of how to achieve the parametric-type interaction between cavity modes and mechanical (mirror or membrane) mode has been discussed in several review papers [27–29]. In what follows, we consider an optomechanical system similar to that considered by Paternostro et al. [23] where entanglement properties between the modes were studied, assuming that only the mirror mode is affected by external thermal fluctuations, i.e., the cavity modes were assumed to be in the ordinary vacuum states. This a a common practice in the study of the dynamics of optomechanical systems to assume that only the oscillating mirror or membrane is in contact with external modes (reservoir), being in a thermal state [30–35]. The ordinary vacuum states of the cavity modes are achieved by the coupling of the modes to an input (external) zero temperature modes. In practice, external modes are not in the ordinary vacuum but rather in non-zero temperature thermal states. Therefore, in what follows, we explore some correlation properties of a three-mode system, illustrated in Figure 1, assuming that the input modes to each of the cavities are in thermal states of unequal mean numbers of photons. The correlation properties of the modes affected by input thermal fields of unequal number of photons is the key point of the present work.

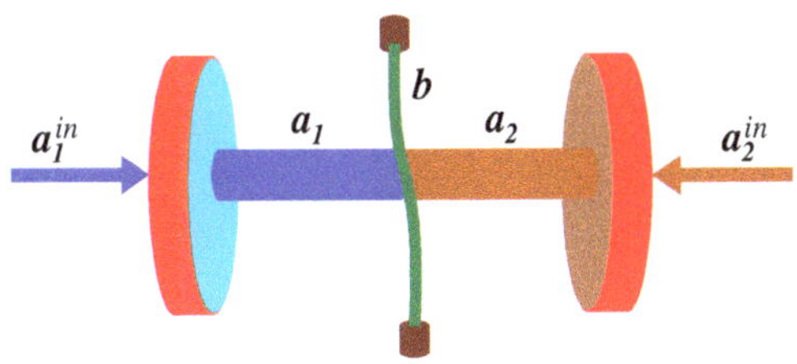

Figure 1. Schematic diagram of the system composed of two single-mode cavities sharing a vibrating membrane. The input fields to the cavities are in thermal states of unequal mean photon numbers.

2.1. Time Evolution of the Modes

We start by writing a complete set of the quantum Langevin equations for the system which can be easily obtained from the Hamiltonian (1) when taking into account dissipation (damping) of the modes and coupling of the modes to external input modes. In the rotating frame, the equations are of the form

$$\begin{aligned} \dot{b} &= -\gamma b + \frac{1}{2} i \left(g_1 a_1^\dagger + g_2 a_2^\dagger \right) + \sqrt{2\gamma}\, b^{in}, \\ \dot{a}_1 &= -\kappa a_1 + \frac{1}{2} i g_1 b^\dagger + \sqrt{2\kappa}\, a_1^{in}, \\ \dot{a}_2 &= -\kappa a_2 + \frac{1}{2} i g_2 b^\dagger + \sqrt{2\kappa}\, a_2^{in}, \end{aligned} \tag{2}$$

where γ is the decay rate of the membrane mode, and we have assumed the same decay rate κ for both cavity modes. Throughout Equation (3), operators a_1^{in}, a_2^{in} and b^{in} are the input noise operators arising from the coupling of the modes to external modes (reservoirs). Here, we assume that the external modes are statistically independent, δ correlated, Gaussian, and in thermal states with

$$\begin{aligned} \langle a_i^{in}(t) a_i^{in\dagger}(t') \rangle &= (n_i + 1)\delta(t - t'), \\ \langle a_i^{in\dagger}(t) a_i^{in}(t') \rangle &= n_i \delta(t - t'), \\ \langle b^{in}(t) b^{in\dagger}(t') \rangle &= (n_b + 1)\delta(t - t'), \\ \langle b^{in\dagger}(t) b^{in}(t') \rangle &= n_b \delta(t - t'), \end{aligned} \tag{3}$$

where $n_i = (\exp\{\hbar\omega/k_B T_i\} - 1)^{-1}$ is the average number of photons in the external modes coupled to the i-th cavity mode of frequency ω and temperature T_i, and $n_b = (\exp\{\hbar\omega/k_B T_b\} - 1)^{-1}$ is the average number of photons in the external modes of temperature T_b coupled to the membrane mode. Thus, in the absence of coupling to the membrane mode the cavity modes, a_1 and a_2 are in thermal states with mean numbers of photons n_1 and n_2, respectively, whereas the membrane is in thermal state with mean number of photons n_b.

2.2. Linear Superpositions of the Modes

It is seen from Equation (3) that mode b interacts simultaneously with both cavity modes. When a mode interacts simultaneously with two other modes, they may act collectively on the given mode. Therefore, it is more convenient to describe the dynamics of the system under consideration in terms of linear superpositions of the cavity modes. Thus, we can transform cavity annihilation operators to linear superpositions a_w and a_u of the form

$$\begin{aligned} a_w &= a_1 \cos\theta + a_2 \sin\theta, \\ a_u &= a_1 \sin\theta - a_2 \cos\theta, \end{aligned} \tag{4}$$

and similarly, for the annihilation operators of the external input fields

$$\begin{aligned} a_w^{in} &= a_1^{in} \cos\theta + a_2^{in} \sin\theta, \\ a_u^{in} &= a_1^{in} \sin\theta - a_2^{in} \cos\theta, \end{aligned} \tag{5}$$

where the mixing angle θ is given by $\tan\theta = g_2/g_1$. Hence in terms of the superposition modes, Equation (3) assumes the simplified form

$$\begin{aligned} \dot{b} &= -\gamma b + \frac{1}{2} i g a_w^\dagger + \sqrt{2\gamma}\, b^{in}, \\ \dot{a}_w &= -\kappa a_w + \frac{1}{2} i g b^\dagger + \sqrt{2\kappa}\, a_w^{in}, \\ \dot{a}_u &= -\kappa a_u + \sqrt{2\kappa}\, a_u^{in}, \end{aligned} \tag{6}$$

where g is the effective coupling strength between the modes, $g = \sqrt{g_1^2 + g_2^2}$.

For both analytical and numerical analyses, it is convenient to write the set of differential Equation (7) in a matrix form

$$\dot{\mathbf{v}} = \mathbf{A}\mathbf{v} + \mathbf{f}_{in}, \tag{7}$$

where $\mathbf{v}^T = [b, a_w^\dagger, a_u]$, $\mathbf{f}_{in}^T = [\sqrt{2\gamma} b^{in}, \sqrt{2\kappa}(a_w^{in})^\dagger, \sqrt{2\kappa} a_u^{in}]$, and the drift matrix $\mathbf{A}$ is given by

$$\mathbf{A} = \begin{pmatrix} -\gamma & \frac{1}{2} i g & 0 \\ -\frac{1}{2} i g & -\kappa & 0 \\ 0 & 0 & -\kappa \end{pmatrix}. \tag{8}$$

From Equation (4) we see that the superpositions of the modes can be controlled through θ by changing the relationship between coupling constants g_1 and g_2. However, the most important property seen from Equation (7) is that the superposition mode determined by the annihilation operator a_u is effectively decoupled from modes a_w and b. On the other hand, the mode a_w is coupled to the membrane mode b with the effective coupling constant g. Despite the lack of the coupling of the a_u mode to the remaining modes, we will show that the mode a_u can exhibit first-order coherence with the mode a_w and the so-called anticoherence with the mode b. The coupling configurations between different modes is shown in Figure 2.

Although the time-dependent solution of Equation (7) is complicated, see Appendix A, the steady-state solution is simple and easily obtained. Therefore, we will focus on the steady-state populations of the modes and correlations between them. We note that the solutions for the populations and correlation functions can be obtained from Equation (7) without approximations by a direct integration of the equations of motion. In the Appendix A, we present a detailed derivation of the steady-state population of the membrane mode.

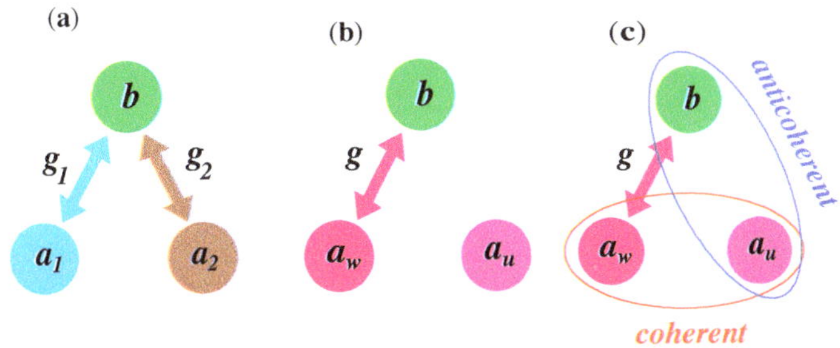

Figure 2. Coupling configurations between modes of the system. (**a**) Couplings between the mirror mode b and the cavity modes a_1 and a_2. (**b**) Couplings between the mode b and the superposition modes a_w and a_u. (**c**) Illustration that the decoupled mode a_u can be coherent with mode a_w and anticoherent with mode b.

3. Populations of the Modes

Let us first examine how different modes are populated in the presence of thermal fields of different mean photon numbers n_i. Solving Equation (7) for the steady-state, we find that the populations of the modes are

$$\begin{aligned}\langle b^{\dagger}b\rangle &= n_b + \frac{\kappa(n_b+1)g^2}{(\kappa+\gamma)(4\kappa\gamma-g^2)} + \frac{\kappa g^2}{(\kappa+\gamma)(4\kappa\gamma-g^2)}(n+\delta n\cos 2\theta),\\ \langle a_w^{\dagger}a_w\rangle &= \frac{\gamma(n_b+1)g^2}{(\kappa+\gamma)(4\kappa\gamma-g^2)} + \left[1+\frac{\gamma g^2}{(\kappa+\gamma)(4\kappa\gamma-g^2)}\right](n+\delta n\cos 2\theta),\\ \langle a_u^{\dagger}a_u\rangle &= n-\delta n\cos 2\theta,\end{aligned} \tag{9}$$

where $n = (n_1+n_2)/2$ is the average number of photons, and $\delta n = (n_1-n_2)/2$ is a difference between the average number of photons in the thermal fields coupled to the cavity modes. Note that δn can vary from $-n$ to $+n$. The populations depend also on the coupling constant g, which cannot be arbitrarily large. The values of g are restricted to those at which the steady-state solutions for the populations are stable, i.e., are positive. It is easily seen from Equation (10) that the positivity of the populations requires $g < \sqrt{4\kappa\gamma}$. Alternatively, conditions for the stability of the steady-state solutions (10) can be determined by applying the Routh–Hurwitz criterion [36] to Equation (7), which says that the components of vector $\mathbf{v}$ decay to stable steady-state values when the determinant of the drift matrix $\mathbf{A}$ is negative. It is easily verified from Equation (8) that det $(\mathbf{A}) < 0$ when $g < \sqrt{4\kappa\gamma}$.

The first important fact we can derive from Equation (10) is that in the case of $\delta n = 0$, the populations depend only on the effective coupling constant g. The difference $\delta n \neq 0$ induces a variation of the populations with the ratio of the coupling constants g_1 and g_2, determined by the mixing angle θ. This means that in the case of $\delta n \neq 0$, by changing the ratio g_2/g_1, i.e., by varying the mixing angle θ, one can change the population of the mode a_u which is decoupled from the remaining modes a_w and b. The transfer rate is proportional to δn, the difference of the thermal occupation of the modes a_1 and a_2. Thus, if only one of the cavity modes is subjected to thermal excitation and the other mode is in a vacuum state, then $\delta n = \pm n$, indicating that the thermal excitation of the cavity mode can be completely and reversibly transferred from modes b and a_w to mode a_u.

The results of our discussion of variations of the populations with θ when the difference $\delta n \neq 0$ are illustrated in Figure 3. We present here variations of the populations with θ for two different values of the effective coupling constant g. As it is seen, for a weak coupling $g \ll \kappa$, the transfer of the population occurs between modes a_w and a_u only. The population of the mode b remains constant. Note the symmetry of the transfer process about $\theta = \pi/4$ corresponding to $g_1 = g_2$. For a strong coupling g, the transfer of the populations between the superposition modes is asymmetric about $\theta = \pi/4\,(g_1 = g_2)$

and is seen to be accompanied by a reduction of the population of mode b. In this case, the population is transferred to mode a_u not only from mode a_w, but also from mode b. Lowering the population of the mode b implies that the system can be employed to cool the mode to a lower temperature. Thus, when $\delta n \neq 0$, it is possible to obtain dramatically reduced populations of the modes. In other words, keeping modes a_1 and a_2 at levels of different thermal occupations ($n_1 \neq n_2$) can work as a mechanism for the cooling of the membrane mode.

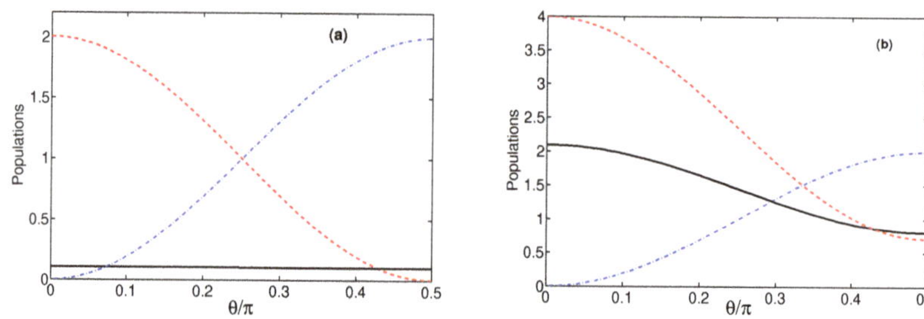

Figure 3. Populations of the modes plotted as a function of θ for $\gamma = \kappa$, $n = 1$, $\delta n = 1$, $n_b = 0.1$ and two different values of the coupling strength g: (**a**) $g = 0.1\kappa$, and (**b**) $g = 1.5\kappa$. Black solid line shows $\langle b^\dagger b\rangle$, dashed red line $\langle a_w^\dagger a_w\rangle$, and dashed-dotted blue line $\langle a_u^\dagger a_u\rangle$.

4. Correlations between the Modes

We now investigate the coherence and correlation effects between the modes when the modes are influenced by thermal fields. We assume that the thermal fields coupled to the cavity modes are of unequal numbers of thermal photons $n_1 \neq n_2$, and the mirror mode is coupled a thermal state with the mean number of phonons n_b.

Different kinds of correlations can exist between the modes. Since the modes are in Gaussian states, which arises from the fact that the Hamiltonian (1) is quadratic, we consider only correlation functions up to a second order only. The correlation functions are expectation values of any combination of operators of two different modes. It is not difficult to show, using Equation (7), that in the steady state, there are the following non-zero correlation functions

$$\begin{aligned}
\langle a_u^\dagger a_w\rangle &= \left[1 + \frac{g^2}{8\kappa(\kappa+\gamma) - g^2}\right]\delta n \sin 2\theta, \\
\langle a_w b\rangle &= \frac{2i\kappa\gamma g}{(\kappa+\gamma)(4\kappa\gamma - g^2)}(n + n_b + 1 + \delta n \cos 2\theta), \\
\langle a_u b\rangle &= \frac{4i\kappa g}{8\kappa(\kappa+\gamma) - g^2}\delta n \sin 2\theta.
\end{aligned} \tag{10}$$

and $\langle a_w^\dagger b\rangle = \langle a_u^\dagger b\rangle = \langle a_w a_u\rangle = 0$. It is seen that the thermal fields of unequal photon numbers $\delta n \neq 0$ induce the first-order coherence between the superposition modes a_u and a_w determined by the function $\langle a_u^\dagger a_w\rangle$, and a correlation between between a_u and b modes determined by the function $\langle a_u b\rangle$, usually called an anomalous correlation function [5,6], or, after Mandel, called anticoherence [19]. As we already mentioned, the nonvanishing correlation function $\langle a_u^\dagger a_w\rangle$ is the signature of the first-order coherence, which may lead to interference effects between the modes. It is well known that the nonvanishing anti-coherence correlation functions $\langle a_w b\rangle$ and $\langle a_u b\rangle$ may lead to entanglement between the involved modes.

It is interesting that the mode a_u which is decoupled from the other modes can exhibit first-order coherence with the mode a_w and anticoherence with mode b. According to Equation (11), this can happen only when $\delta n \neq 0$. To demonstrate this, we examine in detail measures of the degree of coherence and anticoherence.

4.1. Degree of Coherence and Visibility

We already saw that the cross-correlation or mutual coherence function $\langle a_u^\dagger a_w \rangle$ is different from zero when $\delta n \neq 0$. Therefore, the modes can be described as mutually coherent. The degree of coherence of the modes a_u and a_w is defined by the normalized quantity

$$\gamma_{uw}^{(1)} = \frac{|\langle a_u^\dagger a_w \rangle|}{\sqrt{\langle a_u^\dagger a_u \rangle \langle a_w^\dagger a_w \rangle}}, \tag{11}$$

whose values lie between 0 and 1.

In Figure 4, we plot the degree of coherence as a function of δn and θ. Notice that at $\delta n = 0$, the modes are mutually incoherent, regardless of the value of θ. When $\delta n \neq 0$, the modes become mutually coherent. It is clearly seen that for a weak coupling between the modes $g/\kappa \ll 1$, illustrated in Figure 4a, the first-order coherence function is symmetric about $\theta = \pi/4$, and becomes asymmetric when $g/\kappa > 1$, the case corresponding to a strong coupling between modes, illustrated in Figure 4b. In this case, the degree of coherence is reduced in magnitude as θ increases. In the case of a weak coupling, an interesting situation is reached where the coherence attains its maximal value, i.e., the modes become mutually perfectly coherent when $\delta n = n$, i.e., when either n_1 or n_2 is equal to zero. On the other hand, in the strong coupling regime, the degree of coherence is always less than unity.

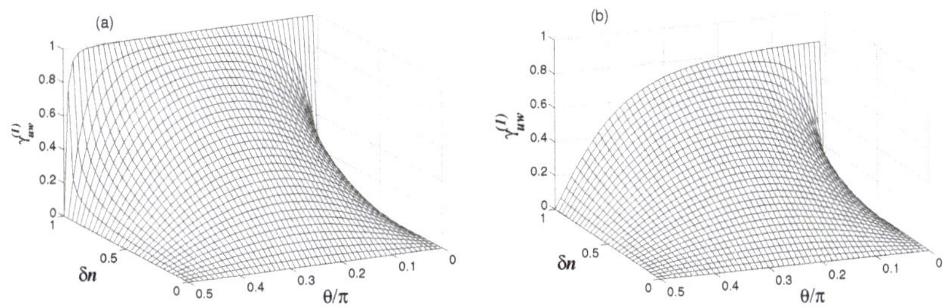

Figure 4. Variation of the degree of coherence between modes a_u and a_w with δn and θ for $\gamma = \kappa$, $n = 1$, $n_b = 0.1$ and two different values of the coupling strength g: (**a**) $g = 0.1\kappa$, and (**b**) $g = 1.5\kappa$.

One can notice from Figure 4a that in the limit of $\delta n = n$, the modes are perfectly coherent when $\theta = 0$, but are completely incoherent when $\theta = \pi/2$. The perfect coherence arises because the definite phase relationship between the modes a_1 and a_2 through the common coupling to the mode b.

Comparing the variation of $\gamma_{uw}^{(1)}$ with the variation of the populations of the modes, shown in Figure 3, we see that $\gamma_{uw}^{(1)}$ can be of equal unity regardless of the distribution of the population between the modes. This surprising behavior has been noticed before in systems of couple parametric downconverters [15,16,37–40], where interference effects were observed between the signal fields of the two downconverters with the degree of coherence $\gamma_{ij}^{(1)} = 1$.

We saw that the modes can be perfectly mutually coherent regardless of the distribution of the population between them. However, the distribution of the population between the modes has an effect on the visibility of the interference pattern and distinguishability of the modes. The visibility $\mathcal{V}$ is determined by the coherence function

$$|\mathcal{V}| = \frac{2|\langle a_u^\dagger a_w \rangle|}{\langle a_u^\dagger a_u \rangle + \langle a_w^\dagger a_w \rangle}, \tag{12}$$

whereas distinguishability is determined by the populations of the modes

$$|\mathcal{D}| = \frac{|\langle a_u^\dagger a_u \rangle - \langle a_w^\dagger a_w \rangle|}{\langle a_u^\dagger a_u \rangle + \langle a_w^\dagger a_w \rangle}, \tag{13}$$

The visibility and distinguishability obey the complementarity relation $|\mathcal{V}|^2 + |\mathcal{D}|^2 \leq 1$, in which the equality holds when the system is described by a pure state. When $|\mathcal{D}| = 0$, the modes are indistinguishable. On the other hand, when $|\mathcal{D}| = 1$, the modes are perfectly distinguished.

The distinguishability $|D|$ is plotted in Figure 5 as a function of δn and θ. For $\delta n = 0$, the distinguishability $|D| = 0$ for all values of θ, indicating that in the case the cavity modes are affected by thermal fields of the same number of photons, and the superposition modes a_w and a_u are undistinguishable independent of the ratio g_2/g_1. For a weak coupling and $\delta n \neq 0$, illustrated in Figure 5a, the distinguishability varies between its minimal value $|D| = 0$ at $\theta = \pi/4$ to its maximal values at $\theta = 0$ and $\theta = \pi/2$. In the completely asymmetric case where $\delta n = \pm n$, the distinguishability $|D| = 1$. More precisely, the modes can be perfectly distinguishable ($|\mathcal{D}| = 1$) only if $\delta n = n$ and either g_1 or g_2 is equal to zero. Thus, in the case of weak and equal coupling constants, $\theta = \pi/4$, the modes are completely non-distinguishable, independent of δn. It is easy to understand if we refer to the fact that in the case of $\theta = \pi/4$, the superpositions a_w and a_u are equally weighted, so that one can not predict from which mode a detected photon came from.

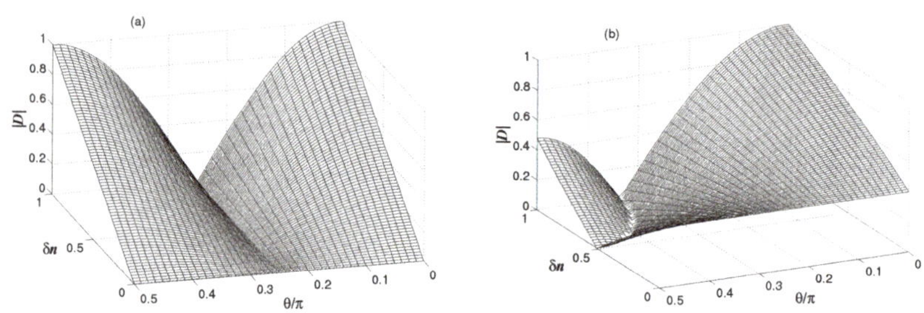

Figure 5. Dependence of the distinguishability $|D|$ on δn and θ for $\gamma = \kappa$, $n = 1$, $n_b = 0.1$ and two different values of the coupling strength g: (**a**) $g = 0.1\kappa$, and (**b**) $g = 1.5\kappa$.

In the case of a strong coupling g, illustrated in Figure 5b, the distinguishability is strongly dependent on the relationship between g_1 and g_2. We see that the modes are always at least partly indistinguishable, except for $\delta n = n$ and $\theta = 0$ at which $|D| = 1$. Moreover, the modes are perfectly indistinguishable at $\theta \neq \pi/4$, i.e., when the modes are coupled to the membrane mode with unequal coupling strengths, $g_1 \neq g_2$.

A close-up view of the variation of the distinguishability $|D|$ with θ at $\delta n = n$ is shown in Figure 6. We also plot the visibility $|V|$ and the complementarity $S = |V|^2 + |D|^2$. The visibility vanishes only when $g_1 = 0$ or $g_2 = 0$, i.e., when one of the cavity modes is decoupled from the membrane mode. In the limit of a weak coupling, $g \ll \kappa$, the visibility and distinguishability are perfectly mutually exclusive, and $S = 1$ for all values of θ, indicating that independent of the ratio g_2/g_1, the system is in a pure state. On the other hand, in the limit of a strong coupling $g > \kappa$, they are no longer perfectly mutually exclusive, i.e., the visibility is greatest for $g_1 \neq g_2$ and the maximum of the visibility does not correspond to the minimum of the distinguishability. Additionally, in this case, $V^2 + |D|^2 < 1$, except $\theta = 0$ at which the modes are perfectly distinguishable. Thus, except $\theta = 0$, the system is in a mixed state. The mixed state results from the fact that in the strong coupling regime, not only the population from mode a_w, but also a population from the membrane mode b is transferred to mode a_u, as it is seen in Figure 3b.

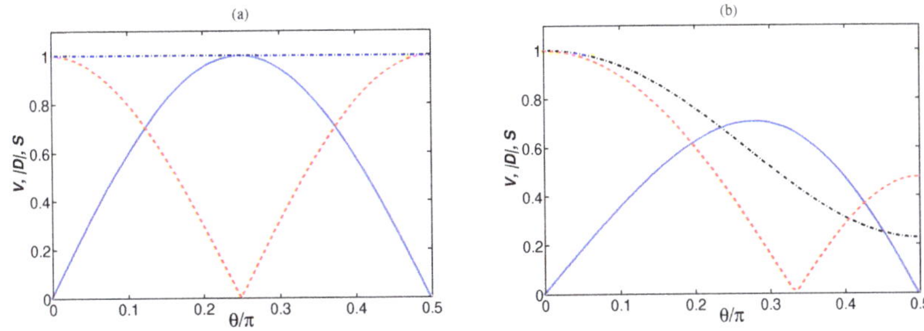

Figure 6. Close-up view of the variation of the distinguishability $|D|$ (red dashed line) with θ at $\delta n = n$ shown in Figure 5 together with the visibility $|V|$ (blue solid line) and complementarity $S = |V|^2 + |D|^2$ (black dashed-dotted line) for $\gamma = \kappa$, $n = 1$, $n_b = 0.1$ and two different values of the coupling g: (**a**) $g = 0.1\kappa$, and (**b**) $g = 1.5\kappa$.

4.2. Degree of Anticoherence and Entanglement

For the uncoupled modes a_u and b, mutual coherence function $\langle a_u^\dagger b\rangle$ is equal zero, and therefore the modes are mutually incoherent. Although the mutual coherence between the modes is equal to zero, it must not be thought that all correlations between the modes are zero. In fact, there are correlations present, but they are reflected by nonzero values of the correlation function $\langle a_u b\rangle$. This happens when $\delta n \neq 0$. Note that $\langle a_u b\rangle \neq 0$ is accompanied by $\langle a_u^\dagger b\rangle = 0$. Following Mandel [19], the correlation function $\langle a_u b\rangle$ is called the anticoherence function, and to quantify the degree of anticoherence, he introduced the measure of anticoherence

$$\gamma_{ub}^{(2)} = \frac{|\langle a_u b\rangle|}{\sqrt{\langle a_u^\dagger b^\dagger a_u b\rangle}}. \tag{14}$$

The values of $\gamma_{ub}^{(2)}$ lie between 0 and 1.

When the modes obey the Gaussian statistics, then [41]

$$\langle a_u^\dagger b^\dagger a_u b\rangle = \langle a_u^\dagger b^\dagger\rangle\langle a_u b\rangle + \langle a_u^\dagger b\rangle\langle b^\dagger a_u\rangle + \langle a_u^\dagger a_u\rangle\langle b^\dagger b\rangle. \tag{15}$$

Since $\langle a_u^\dagger b\rangle = 0$, Equation (14) then gives

$$\gamma_{ub}^{(2)} = \frac{\eta_{ub}}{\sqrt{\eta_{ub}^2 + 1}}, \tag{16}$$

where

$$\eta_{ub} = \frac{|\langle a_u b\rangle|}{\sqrt{\langle a_u^\dagger a_u\rangle\langle b^\dagger b\rangle}}. \tag{17}$$

is the normalized anomalous correlation function. Thus, the

The nonvanishing anticoherence corresponds to a situation in which the modes could be entangled. In order to connect anticoherence to entanglement, we consider the Cauchy–Schwarz inequality, which is often used to identify entanglement [3]. The Cauchy–Schwarz inequality for the modes a_u and b is verified by reference to the so-called Cauchy–Schwartz parameter χ_{ub} involving the second-order correlation functions

$$\chi_{ub} = \frac{g_u^{(2)} g_b^{(2)}}{\left(g_{ub}^{(2)}\right)^2}, \tag{18}$$

where

$$g_{ub}^{(2)} = \frac{\langle a_u^\dagger b^\dagger a_u b\rangle}{\langle a_u^\dagger a_u\rangle\langle b^\dagger b\rangle} \tag{19}$$

is the normalized second-order cross correlation function, and

$$g_u^{(2)} = \frac{\langle a_u^{\dagger 2} a_u^2\rangle}{\langle a_u^\dagger a_u\rangle^2}, \quad g_b^{(2)} = \frac{\langle b^{\dagger 2} b^2\rangle}{\langle b^\dagger b\rangle^2}, \tag{20}$$

are the normalized intensity autocorrelation functions of the modes a_u and b, respectively.

Using the Gaussian-mode decomposition (15), the correlation functions can be readily related to the coherence functions

$$\begin{aligned} g_i^{(2)} &= 2+\eta_{ii}^2, \quad i = u, b, \\ g_{ub}^{(2)} &= 1+\left(\gamma_{ub}^{(1)}\right)^2+\eta_{ub}^2. \end{aligned} \tag{21}$$

Since in our case, $\eta_{uu} = \eta_{bb} = \gamma_{ub}^{(1)} = 0$, the Cauchy–Schwarz parameter takes the form

$$\chi_{ub} = \frac{4}{(1+\eta_{ub}^2)^2}, \tag{22}$$

which can be expressed in terms of the degree of the anticoherence as

$$\chi_{ub} = 4\left[1-\left(\gamma_{ub}^{(2)}\right)^2\right]^2. \tag{23}$$

To examine the occurrence of entanglement, we must check whether the Cauchy–Schwarz inequality ($\chi_{ub} > 1$) is violated. From Equation (23), we see that the condition that the modes are anticoherent, i.e., $\gamma_{ub}^{(2)}$ is a necessary but not sufficient condition for entanglement between the modes. In other words, the modes could be anticoherent but not enough to obtain $\chi_{ub} < 1$. It is easily verified that for the Cauchy–Schwarz inequality to be violated, it is necessary that $\gamma_{ub}^{(2)} > 1/\sqrt{2}$. Thus, for two modes to be entangled, they should be anticoherent to a degree about 71%.

Figure 7a shows the Cauchy–Schwarz parameter χ_{ub} as a function of δn and θ. It is clearly seen that the parameter χ_{ub} is reduced below its maximal value $\chi_{ub} = 4$ when $\delta n \neq 0$. The parameter χ_{ub} decreases to a minimum value at $\delta n = n$, but unfortunately the minimal value is not smaller than the threshold for entanglement ($\chi_{ub} = 1$). This indicates that the anticoherence between the modes is not strong enough for the modes a_u and b to be entangled.

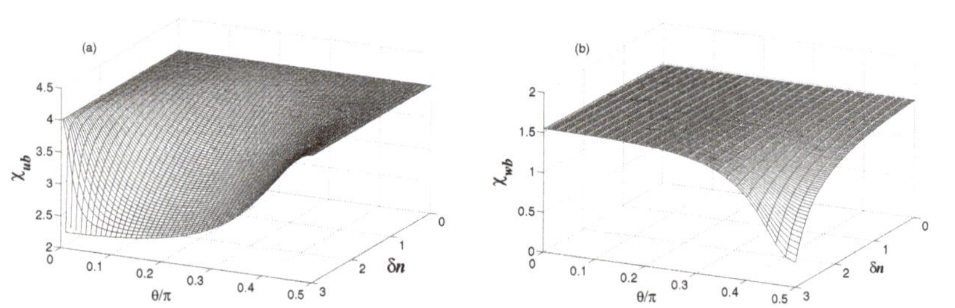

Figure 7. Variation of the Cauchy–Schwarz parameters (**a**) χ_{ub} and (**b**) χ_{wb} with δn and θ for $\gamma = \kappa$, $n = 3$, $n_b = 0.1$ and $g = \kappa$.

Although modes a_u and b are not entangled, there could be entanglement between modes a_w and b, which are directly coupled to each other. The results for the Cauchy–Schwarz parameter χ_{wb} are shown in Figure 7b. It is seen that for certain values of δn and θ, the parameter χ_{wb} can be reduced below the threshold for entanglement. It was noticed before that in the case when the cavity modes are affected by thermal fields of the same photon numbers $(n_1 = n_2 = n)$, entanglement between cavity mode and the membrane mode is restricted to very small values of $n < 1/2$. The results shown in Figure 7b are in sharp contrast to the case of equal number of thermal excitations, where entanglement is restricted to very small values of n and indicate quite clearly that in the case of unequal photon numbers $(n_1 \neq n_2)$, entanglement between the modes can be observed, even for large values of n.

In physical terms, we may attribute the appearance of entanglement between modes a_w and b when $n_1 \neq n_2$ to the fact that a part of the population of the modes, which has a destructive effect on entanglement, is transferred and stored in the decoupled mode a_u.

Before concluding, we note that although we have discussed and graphically illustrated the coherence and anticoherence properties of the modes only for the case of equal damping rates of the modes, $\gamma = \kappa$, analogous results are obtained in the experimentally realistic case of $\gamma \ll \kappa$ [27–29].

As an illustration, in Figure 8, we plot $\gamma_{uw}^{(1)}$ and χ_{wb} for $\gamma = 0.01\kappa$. Comparing the results with those presented in Figures 4b and 7b we saw that $\gamma_{uw}^{(1)}$ and χ_{wb} behave in qualitatively the same manner as for $\gamma = \kappa$. While the maximal value of the coherence $\gamma_{uw}^{(1)}$ between uncoupled modes is reduced for $\gamma \ll \kappa$ compared with Figure 4b, it is still nonzero over the entire range of $\delta n \neq 0$. Similarly, although the parameter χ_{wb} has risen for $\gamma \ll \kappa$ compared with Figure 7b, the region near $\delta n = n$ still shows reduction of χ_{wb} below the threshold for entanglement.

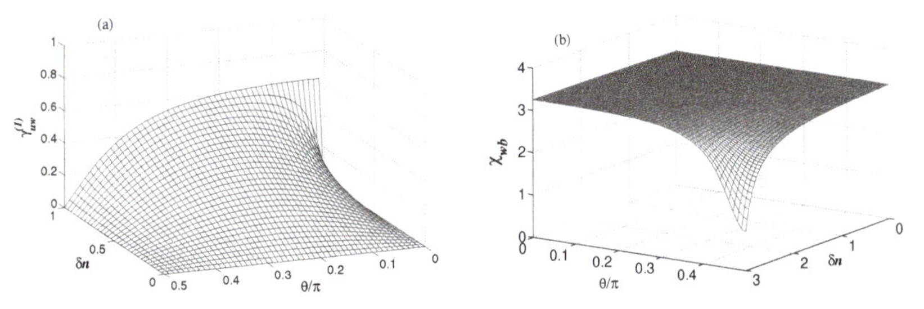

Figure 8. (**a**) Variation of the degree of coherence $\gamma_{uw}^{(1)}$ with δn and θ for $\gamma = 0.01\kappa$, $n = 1$, $n_b = 0.1$ and $g = 0.19\kappa$. (**b**) Variation of the Cauchy-Schwarz parameter χ_{wb} with δn and θ for $\gamma = 0.01\kappa$, $n = 3$, $n_b = 0.1$ and $g = 0.19\kappa$.

5. Conclusions

We considered coherence properties between modes of a three-mode optomechanical system composed of two cavity modes simultaneously coupled to a membrane mode. We obtained analytical solutions for the steady-state populations of the modes and correlation functions describing coherence effects between the modes. Working in terms of linear superpositions of the cavity modes, we showed that one of the linear superpositions can be completely decoupled from the remaining modes. In spite of this, we found that the decoupled superposition can be completely coherent with the other superposition modes and can simultaneously exhibit anticoherence with the membrane mode. A detailed analysis showed that these correlation effects can happen only when the cavity modes are affected by the external input modes being in thermal states of unequal average photon numbers. In particular, we found that the coherences have a substantial effect on population distribution between the modes such that the population can be reversibly transferred between the superposition modes. The transfer of the population can lead to lowering of the population of the membrane mode. Therefore, the system can be considered as an alternative way to cool modes to lower temperatures. We also showed that a difference of the average numbers of photons in the thermal fields may affect entanglement between the superposition mode directly coupled to the membrane mode such that it may occur in a less restricted range of the number of thermal photons. In other words, the modes could be entangled, even with large numbers of thermal photons.

Author Contributions: Conceptualization, L.S., W.Y. and Z.F.; methodology, L.S. and Z.F.; software, Y.L., C.L. and K.Z.; validation, Y.L., C.L. and K.Z.; formal analysis, L.S., W.Y. and Z.F.; investigation, L.S., Y.L. and C.L.; writing—original draft preparation, Z.F.; writing—review and editing, L.S. and Z.F. All authors have read and agreed to the published version of the manuscript.

Funding: This work was supported by National Science Foundation (NSF) of China (Grant Nos. 11374050, 11774054, and 12075036).

Institutional Review Board Statement: Not applicable.

Informed Consent Statement: Not applicable.

Data Availability Statement: Not applicable.

Conflicts of Interest: The authors declare no conflict of interest.

Appendix A. Evaluation of the Steady-State Population of the Membrane Mode

In this Appendix, we provide some details of the derivation of the analytical expression for the steady-state population of the mode b. Using Equation (7), we find that a former integration of the equations for b and a_w leads to

$$b(t) = b(0)e^{-\gamma t} + \frac{1}{2}ige^{-\gamma t}\int_0^t dt' a_w^\dagger(t')e^{\gamma t'} + \sqrt{2\gamma}e^{-\gamma t}\int_0^t dt' b^{in}(t')e^{\gamma t'}, \tag{A1}$$

$$a_w(t) = a_w(0)e^{-\kappa t} + \frac{1}{2}ige^{-\kappa t}\int_0^t dt' b^\dagger(t')e^{\kappa t'} + \sqrt{2\kappa}\,e^{-\kappa t}\int_0^t dt' a_w^{in}(t')e^{\kappa t'}. \tag{A2}$$

Substituting the expression for $a_w^\dagger$ into Equation (A1) and using the double integration rule

$$\int_0^t dt' A(t')\int_0^{t'} dt'' B(t'') = \int_0^{t} dt' B(t')\int_{t'}^t dt'' A(t'') \tag{A3}$$

we find that the expression for $b(t)$ can be written as

$$b(t) = y(t) + \int_0^t dt' K(t,t')b(t'), \tag{A4}$$

where $K(t,t')$ is the kernel of the integral of the form

$$K(t,t') = \frac{g^2}{4(\gamma-\kappa)}\left(e^{-\kappa(t-t')} - e^{-\gamma(t-t')}\right), \tag{A5}$$

and the term $y(t)$ has the form

$$\begin{aligned} y(t) &= b(0)e^{-\gamma t} - \tfrac{g a_w^\dagger(0)}{2(\gamma-\kappa)}\left[e^{-\kappa t} - e^{-\gamma t}\right] + \sqrt{2\gamma}\textstyle\int_0^t dt' b^{in\dagger}(t')e^{-\gamma(t-t')} \\ &+ i\tfrac{\sqrt{2\kappa}g}{2(\gamma-\kappa)}\textstyle\int_0^t dt' a_w^{in\dagger}(t')\left[e^{-\kappa(t-t')} - e^{-\gamma(t-t')}\right]. \end{aligned} \tag{A6}$$

It is seen that the kernel $K(t,t')$ depends only on the time difference $t-t'$, and may be written in the form

$$K(t,t') = \frac{g^2}{4(\gamma-\kappa)}H(t,t') = \lambda H(t,t'), \tag{A7}$$

where $\lambda = g^2/4(\gamma - \kappa)$.

The integral Equation (A4) can be solved using the Laplace transformation. Thus if

$$\begin{aligned} \int_0^t dt' H(t') e^{-pt'} &= H(p), \\ \int_0^t dt' y(t') e^{-pt'} &= y(p), \\ \int_0^t dt' b(t') e^{-pt'} &= b(p), \end{aligned} \tag{A8}$$

we obtain from Equation (A4)

$$b(p) = \frac{y(p)}{1 - \lambda\, H(p)}, \tag{A9}$$

where

$$H(p) = \frac{1}{p+\kappa} - \frac{1}{p+\gamma}, \tag{A10}$$

and $y(p)$ is

$$\begin{aligned} y(p) &= \left\{ b(0) + \sqrt{2\gamma} B(p) + \tfrac{g}{2(\gamma-\kappa)} \left[a_w^\dagger(0) - i\sqrt{2\kappa} A_w^\dagger(p) \right] \right\} \tfrac{1}{p+\gamma} \\ &- \tfrac{g}{2(\gamma-\kappa)} \left[a_w^\dagger(0) + i\sqrt{2\kappa} A_w^\dagger(p) \right] \tfrac{1}{p+\kappa}, \end{aligned} \tag{A11}$$

with

$$\begin{aligned} B(p) &= \int_0^t dt' b^{in}(t') e^{-pt'}, \\ A^\dagger(p) &= \int_0^t dt' a_w^{in\dagger}(t') e^{-pt'}. \end{aligned} \tag{A12}$$

Substituting the solution (A10) for $H(p)$ into Equation (A9), we readily find

$$b(p) = \frac{y(p)(p+\kappa)(p+\gamma)}{(p+\kappa)(p+\gamma) - \lambda\,(\gamma-\kappa)}. \tag{A13}$$

Having available the Laplace transform $b(p)$, we find $b(t)$ simply by taking the inverse of the Laplace transformation (A13). We then obtain

$$\begin{aligned} b(t) &= \textstyle\sum_{i=1}^{2} (p - p_i) b(p_i) e^{p_i t} \\ &= b(0) \left[\tfrac{(\kappa-\gamma)}{\Delta} \sinh\left(\tfrac{1}{2}\Delta t\right) + \cosh\left(\tfrac{1}{2}\Delta t\right) \right] \mathrm{e}^{-\frac{1}{2}(\kappa+\gamma)t} + a_w^\dagger(0) \tfrac{ig}{\Delta} \sinh\left(\tfrac{1}{2}\Delta t\right) e^{-\frac{1}{2}(\kappa+\gamma)t} \\ &+ \tfrac{ig\sqrt{2\kappa}}{2\Delta} \left[A^\dagger(p_1) e^{\frac{1}{2}\Delta t} - A^\dagger(p_2) e^{-\frac{1}{2}\Delta t} \right] e^{-\frac{1}{2}(\kappa+\gamma)t} \\ &+ \tfrac{\sqrt{2\gamma}}{2\Delta} \left\{ [(\kappa-\gamma)+\Delta] B(p_1) e^{\frac{1}{2}\Delta t} - [(\kappa-\gamma)-\Delta] B(p_2) e^{-\frac{1}{2}\Delta t} \right\} e^{-\frac{1}{2}(\kappa+\gamma)t}, \end{aligned} \tag{A14}$$

where

$$p_{1,2} = \frac{1}{2}(\kappa+\gamma) \pm \frac{1}{2}\sqrt{(\kappa-\gamma)^2 + g^2} \tag{A15}$$

are roots of the quadratic equation

$$(p+\kappa)(p+\gamma) - \frac{1}{4}g^2 = 0, \tag{A16}$$

and $\Delta = \sqrt{(\kappa-\gamma)^2 + g^2}$.

We can use the solution $b(t)$ to find the population of the mode b simply multiplying $b(t)$ from the left by $b^\dagger(t)$ and then taking the expectation value. We thus find

$$
\begin{aligned}
\langle b^{\dagger}(t)b(t)\rangle &= \langle b^{\dagger}(0)b(0)\rangle\left[\tfrac{(\kappa-\gamma)}{\Delta}\sinh\left(\tfrac{1}{2}\Delta t\right)+\cosh\left(\tfrac{1}{2}\Delta t\right)\right]^{2}e^{-(\kappa+\gamma)t} \\
&+ \left[\langle a_w^{\dagger}(0)a_w(0)\rangle+1\right]\tfrac{g^2}{\Delta^2}\sinh^2\left(\tfrac{1}{2}\Delta t\right)e^{-(\kappa+\gamma)t} \\
&+ \tfrac{\kappa g^2}{2\Delta^2}\left[\langle A(p_1)A^{\dagger}(p_1)\rangle e^{\Delta t}+\langle A(p_2)A^{\dagger}(p_2)\rangle e^{-\Delta t}\right. \\
&- \left.\langle A(p_1)A^{\dagger}(p_2)\rangle-\langle A(p_2)A^{\dagger}(p_1)\rangle\right]e^{-(\kappa+\gamma)t} \\
&+ \tfrac{\gamma}{2\Delta^2}\{[(\kappa-\gamma)+\Delta]^2\langle B^{\dagger}(p_1)B(p_1)\rangle e^{\Delta t}+[(\kappa-\gamma)-\Delta]^2\langle B^{\dagger}(p_2)B(p_2)\rangle e^{-\Delta t} \\
&+ g^2(\langle B^{\dagger}(p_1)B(p_2)\rangle+\langle B^{\dagger}(p_2)B(p_1)\rangle)\}e^{-(\kappa+\gamma)t},
\end{aligned} \tag{A17}
$$

where

$$
\langle A(p_i)A^{\dagger}(p_i)\rangle=\int_0^t dt'\int_0^t dt''\langle a_w^{in}(t')a_w^{in\dagger}(t'')\rangle e^{-p_i(t'+t'')},\ i=1,2, \tag{A18}
$$

and

$$
\langle B^{\dagger}(p_i)B(p_i)\rangle=\int_0^t dt'\int_0^t dt''\langle b^{in\dagger}(t')b^{in}(t'')\rangle e^{-p_i(t'+t'')},\ i=1,2. \tag{A19}
$$

Since $\langle b^{in\dagger}(t')b^{in}(t'')\rangle=n_b\delta(t'-t'')$, we get

$$
\langle B^{\dagger}(p_i)B(p_i)\rangle=n_b\int_0^t dt'e^{-2p_it'}=\frac{n_b}{2p_i}\left(1-e^{-2p_it}\right),\ i=1,2, \tag{A20}
$$

and

$$
\langle B^{\dagger}(p_1)B(p_2)\rangle=\langle B^{\dagger}(p_2)B(p_1)\rangle=\frac{n_b}{p_1+p_2}\left[1-e^{-(p_1+p_2)t}\right]. \tag{A21}
$$

Similarly, since

$$
\langle a_w^{in}(t')a_w^{in\dagger}(t'')\rangle=(n_w+1)\delta(t'-t''), \tag{A22}
$$

where $n_w=(g_1^2n_1+g_2^2n_2)/g^2$, we get

$$
\langle A(p_1)A^{\dagger}(p_1)\rangle=(n_w+1)\int_0^t dt'e^{-2p_1t'}=\frac{(n_w+1)}{(\kappa+\gamma)-\Delta}\left(e^{(\kappa+\gamma-\Delta)t}-1\right), \tag{A23}
$$

and

$$
\begin{aligned}
\langle A(p_2)A^{\dagger}(p_2)\rangle &= \frac{(n_w+1)}{(\kappa+\gamma)+\Delta}\left(e^{(\kappa+\gamma+\Delta)t}-1\right), \\
\langle A(p_1)A^{\dagger}(p_2)\rangle &= \langle A(p_2)A^{\dagger}(p_1)\rangle=\frac{(n_w+1)}{(\kappa+\gamma)}\left(e^{(\kappa+\gamma)t}-1\right).
\end{aligned} \tag{A24}
$$

Substituting these results for the correlation functions into Equation (A17), and taking the limit of $t\to\infty$, we obtain

$$
\lim_{t\to\infty}\langle b^{\dagger}(t)b(t)\rangle=n_b+\frac{\kappa(n_b+1)g^2}{(\kappa+\gamma)(4\kappa\gamma-g^2)}+\frac{\kappa g^2}{(\kappa+\gamma)(4\kappa\gamma-g^2)}n_w. \tag{A25}
$$

Writing n_w in terms of $n=(n_1+n_2)/2$, $\delta n=(n_1-n_2)/2$, and $\tan\theta=g_2/g_1$, we obtain the expression for the population of the mode b given in Equation (10).

References

1. Glauber, R.J. Photon correlations. *Phys. Rev. Lett.* **1963**, *10*, 84. [CrossRef]
2. Glauber, R.J. The quantum theory of optical coherence. *Phys. Rev.* **1963**, *130*, 2529. [CrossRef]
3. Mandel, L.; Wolf, E. *Optical Coherence and Quantum Optics*; Cambridge University Press: Cambridge, UK, 1995.

4. Ficek, Z.; Swain, S. *Quantum Interference and Coherence: Theory and Experiments*; Springer: New York, NY, USA, 2004.
5. Agarwal, G.S. Anomalous coherence functions of the radiation fields. *Phys. Rev. A* **1986**, *33*, 11584. [CrossRef] [PubMed]
6. Heidmann, A.; Reynaud, S. Squeezing and antibunching in phase-matched many-atom resonance fluorescence. *J. Mod. Opt.* **1987**, *34*, 923. [CrossRef]
7. Wolf, S.; Wechs, J.; von Zanthier, J.; Schmidt-Kaler, F. Visibility of Young's interference fringes: Scattered light from small ion crystals. *Phys. Rev. Lett.* **2016**, *116*, 183002. [CrossRef]
8. Obsil, P.; Lesundák, A.; Pham, T.; Araneda, G.; Cizek, M.; Cip, O.; Filip, R.; Slodicka, L. Multipath interference from large trapped ion chains. *New J. Phys.* **2019**, *21*, 093039. [CrossRef]
9. Mandel, L. Photon interference and correlation effects produced by independent quantum sources. *Phys. Rev. A* **1983**, *28*, 929. [CrossRef]
10. Ghosh, R.; Hong, C.K.; Ou, Z.Y.; Mandel, L. Interference of two photons in parametric down conversion. *Phys. Rev. A* **1986**, *34*, 3962. [CrossRef]
11. Silverstone, J.W.; Bonneau, D.; Ohira, K.; Suzuki, N.; Yoshida, H.; Iizuka, N.; Ezaki, M.; Natarajan, C.M.; Tanner, M.G.; Hadfield, R.H.; et al. On-chip quantum interference between silicon photon-pair sources. *Nat. Photonics* **2014**, *8*, 104. [CrossRef]
12. Preble, S.F.; Fanto, M.L.; Steidle, J.A.; Tison, C.C.; Howland, G.A.; Wang, Z.; Alsing, P.M. On-chip quantum interference from a single silicon ring-resonator source. *Phys. Rev. Appl.* **2015**, *4*, 021001. [CrossRef]
13. Barnett, S.M.; Knight, P.L. Squeezing in correlated quantum systems. *J. Mod. Opt.* **1987**, *34*, 841. [CrossRef]
14. Ficek, Z.; Tanaś, R. *Quantum-Limit Spectroscopy*; Springer: New York, NY, USA, 2017.
15. Wang, L.J.; Zou, X.Y.; Mandel, L. Induced coherence without induced emission. *Phys. Rev. A* **1991**, *44*, 4614. [CrossRef] [PubMed]
16. Heuer, A.; Menzel, R.; Milonni, P.W. Induced coherence, vacumm fields, and complementatity in biphoton generation. *Phys. Rev. Lett.* **2015**, *114*, 053601. [CrossRef] [PubMed]
17. Ou, Z.Y.; Mandel, L. Further evidence of nonclassical behavior in optical interference. *Phys. Rev. Lett.* **1989**, *62*, 2941. [CrossRef]
18. Rubin, M.H.; Klyshko, D.N.; Shih, Y.H.; Sergienko, A.V. Theory of two-photon entanglement in type-II optical parametric down-conversion. *Phys. Rev. A* **1994**, *50*, 5122. [CrossRef]
19. Mandel, L. Anticoherence. *Pure Appl. Opt.* **1998**, *7*, 927. [CrossRef]
20. Armstrong, S.; Wang, M.; Teh, R.Y.; Gong, Q.H.; He, Q.Y.; Janousek, J.; Bachor, H.A.; Reid, M.D.; Lam, P.K. Multipartite Einstein-Podolsky-Rosen steering and genuine tripartite entanglement with optical networks. *Nat. Phys.* **2015**, *11*, 167. [CrossRef]
21. Parkins, A.S.; Solano, E.; Cirac, J.I. Unconditional two-mode squeezing of separated atomic ensembles. *Phys. Rev. Lett.* **2006**, *96*, 053602. [CrossRef]
22. Sun, L.H.; Li, G.X.; Gu, W.J.; Ficek, Z. Generating coherence and entanglement with a finite-size atomic ensemble in a ring cavity. *New J. Phys.* **2011**, *13*, 093019. [CrossRef]
23. Paternostro, M.; Vitali, D.; Gigan, S.; Kim, M.S.; Brukner, C.; Eisert, J.; Aspelmeyer, M. Creating and probing multipartite macroscopic entanglement with light. *Phys. Rev. Lett.* **2007**, *99*, 250401. [CrossRef]
24. Shkarin, A.B.; Flowers-Jacobs, N.E.; Hoch, S.W.; Kashkanova, A.D.; Deutsch, C.; Reichel, J.; Harris, J.G.E. Optically mediated hybridization between two mechanical modes. *Phys. Rev. Lett.* **2014**, *112*, 013602. [CrossRef] [PubMed]
25. Xu, H.H.; Jiang, L.; Clerk, A.A.; Harris, J.G.E. Nonreciprocal control and cooling of phonon modes in an optomechanical system. *Nature* **2019**, *568*, 65. [CrossRef] [PubMed]
26. Heinrich, G.; Ludwig, M.; Wu, H.; Hammerer, K.; Marquardt, F. Dynamics of coupled multimode and hybrid optomechanical systems. *C. R. Phys.* **2011**, *12*, 837. [CrossRef]
27. Genes, C.; Mari, A.; Vitali, D.; Tombesi, P. Quantum effects in optomechanical systems. *Adv. At. Mol. Opt. Phys.* **2009**, *57*, 33.
28. Meystre, P. A short walk through quantum optomechanics. *Ann. Phys.* **2013**, *525*, 215. [CrossRef]
29. Aspelmeyer, M.; Kippenberg, T.; Marquardt, F. Cavity optomechanics. *Rev. Mod. Phys.* **2014**, *86*, 1391. [CrossRef]
30. Vitali, D.; Tombesi, P.; Woolley, M.J.; Doherty, A.C.; Milburn, G.J. Entangling a nanomechanical resonator and a superconducting microwave cavity. *Phys. Rev. A* **2007**, *76*, 042336. [CrossRef]
31. Aspelmeyer, M.; Gröblacher, S.; Hammerer, K.; Kiesel, N. Quantum optomechanics—Throwing a glance. *J. Opt. Soc. Am. B* **2010**, *27*, A189–A197. [CrossRef]
32. Hofer, S.G.; Wieczorek, W.; Aspelmeyer, M.; Hammerer, K. Quantum entanglement and teleportation in pulsed cavity optomechanics. *Phys. Rev. A* **2011**, *84*, 052327. [CrossRef]
33. Sun, L.H.; Li, G.X.; Ficek, Z. First-order coherence versus entanglement in a nanomechanical cavity. *Phys. Rev. A* **2012**, *85*, 022327. [CrossRef]
34. Sun, F.X.; Mao, D.; Dai, Y.T.; Ficek, Z.; He, Q.Y.; Gong, Q.H. Phase control of entanglement and quantum steering in a three-mode optomechanical system. *New J. Phys.* **2017**, *19*, 123039. [CrossRef]
35. de Lépinay, L.M.; Ockeloen-Korppi, C.F.; Woolley, M.J.; Sillanpää, M.A. Quantum mechanics–free subsystem with mechanical oscillators. *Science* **2021**, *372*, 625. [CrossRef] [PubMed]
36. Gradshteyn, I.S.; Ryzhik, I.M. *Table of Integrals, Series and Products*; Academic Press: Orlando, FL, USA, 1980; p. 1119.
37. Heuer, A.; Menzel, R.; Milonni, P.W. Complementarity in biphoton generation with stimulated or induced coherence. *Phys. Rev. A* **2015**, *92*, 033834. [CrossRef]
38. Menzel, R.; Heuer, A.; Milonni, P.W. Entanglement, complementarity, and vacuum fields in spontaneous parametric down-conversion. *Atoms* **2019**, *7*, 27. [CrossRef]

39. Lahiri, M.; Hochrainer, A.; Lapkiewicz, R.; Lemos, G.B.; Zeilinger, A. Nonclassicality of induced coherence without induced emission. *Phys. Rev. A* **2019**, *100*, 053839. [CrossRef]
40. Wiseman, H.M.; Molmer, K. Induced coherence with and without induced emission. *Phys. Lett. A* **2000**, *270*, 245. [CrossRef]
41. Gardiner, C.W.; Zoller, P. *Quantum Noise*; Springer: New York, NY, USA, 2000; p. 122.

entropy

Article

The Generalized Euler Characteristics of the Graphs Split at Vertices

Omer Farooq *, Michał Ławniczak *, Afshin Akhshani, Szymon Bauch and Leszek Sirko *

Institute of Physics, Polish Academy of Sciences, Aleja Lotników 32/46, 02-668 Warsaw, Poland; akhshani@ifpan.edu.pl (A.A.); bauch@ifpan.edu.pl (S.B.)
* Correspondence: omer.farooq@ifpan.edu.pl (O.F.); lawni@ifpan.edu.pl (M.Ł.); sirko@ifpan.edu.pl (L.S.)

Abstract: We show that there is a relationship between the generalized Euler characteristic $\mathcal{E}_o(|V_{D_o}|)$ of the original graph that was split at vertices into two disconnected subgraphs $i = 1, 2$ and their generalized Euler characteristics $\mathcal{E}_i(|V_{D_i}|)$. Here, $|V_{D_o}|$ and $|V_{D_i}|$ denote the numbers of vertices with the Dirichlet boundary conditions in the graphs. The theoretical results are experimentally verified using microwave networks that simulate quantum graphs. We demonstrate that the evaluation of the generalized Euler characteristics $\mathcal{E}_o(|V_{D_o}|)$ and $\mathcal{E}_i(|V_{D_i}|)$ allow us to determine the number of vertices where the two subgraphs were initially connected.

Keywords: quantum graphs; microwave networks; Euler characteristic; Neumann and Dirichlet boundary conditions

Citation: Farooq, O.; Ławniczak, M.; Akhshani, A.; Bauch, S.; Sirko, L. The Generalized Euler Characteristics of the Graphs Split at Vertices. *Entropy* **2022**, *24*, 387. https://doi.org/10.3390/e24030387

Academic Editor: Adam Gadomski

Received: 2 February 2022
Accepted: 8 March 2022
Published: 9 March 2022

1. Introduction

The concept of graphs was already introduced in the XVIII century by Leonhard Euler [1]. Two hundred years later, Linus Pauling [2] considered quantum graphs in order to describe the motion of quantum particles in a physical network. The models of quantum graphs were widely used to investigate many physical systems, e.g., quantum wires [3], mesoscopic quantum systems [4,5], a topological edge invariant [6], and the photon number statistics of coherent light [7]. Broad applications of graphs and networks mean that the theory of quantum graphs has been a subject of extensive research [8–14].

We will consider a metric graph $\Gamma = (V, E)$, which consists of v vertices, $v \in V$, connected by e edges, $e \in E$. The edges e are intervals of the length l_e on the real line $\mathbb{R}$. The metric graph becomes quantum when we equip it with the free Schrödinger operator. In our case, this is the one-dimensional Laplace operator, which equals $L(\Gamma) = -\frac{d^2}{dx_e^2}$ on each of the edges $e \in E$ of the graph Γ. The self-adjoint Laplace operator $L(\Gamma)$ has a discrete and non-negative spectrum [12].

A signal inside a graph moves along the edges, and at each vertex $v \in V$ it splits and enters all edges adjacent to v. If the signal enters the vertex v along the edge e' and leaves it along the edge e, then the ratio of amplitudes of entering and leaving signals is given by the vertex scattering matrix, which depends on the vertex boundary condition. We will consider two types of vertex boundary conditions. The standard boundary conditions are called also Neumann boundary conditions, for which the eigenfunctions are continuous at vertices and the sums of their oriented derivatives at vertices are zero. The vertex scattering matrix corresponding to the Neumann boundary conditions [15] is given by

$$ {}^{N}\sigma_{e,e'}^{(v)} = \frac{2}{d_v} - \delta_{e,e'}, \tag{1} $$

where d_v is the degree of the vertex v, i.e., the number of edges incident to the vertex v, and $\delta_{e,e'}$ is the Kronecker delta. The vertices with the Neumann boundary conditions will be denoted as v_N.

For the Dirichlet boundary condition, an eigenfunction at the vertex takes the value zero, which leads to the vertex scattering matrix [15,16]

$$^{D}\sigma^{(v)}_{e,e'} = -\delta_{e,e'}. \tag{2}$$

One should point out that the Dirichlet boundary conditions are imposed only at degree one vertices and higher-degree Dirichlet vertices should be treated as separate degree one Dirichlet vertices. The vertices with the Dirichlet boundary conditions will be denoted as v_D. Different types of the boundary conditions, including the Neumann and Dirichlet ones for higher-dimensional systems such as grains, are comprehensively described in Refs. [17,18].

The total number of vertices $|V|$ in a general graph, consisting of both Neumann and Dirichlet boundary conditions, is defined by $|V| = |V_N| + |V_D|$, where $|V_N|$ and $|V_D|$ denote the number of vertices with Neumann and Dirichlet boundary conditions, respectively.

One of the most important characteristics of metric graphs $\Gamma = (V, E)$ with the standard boundary conditions ($|V_D| = 0$) is the Euler characteristic

$$\chi = |V| - |E|, \tag{3}$$

where $|V|$ and $|E|$ denote the number of vertices and edges of the graph. It is a purely topological quantity; however, it has been shown in [19–22] that it can also be defined by the graph and microwave network spectra. The formula describing the generalized Euler characteristic $\mathcal{E}$ [22,23], which is also applicable for graphs and networks with the Dirichlet boundary conditions, will be discussed later.

In the experimental investigation of properties of quantum graphs, we used microwave networks simulating quantum graphs [16,24–29]. The emulation of quantum graphs by microwave networks is possible because of the formal analogy of the one-dimensional Schrödinger equation describing quantum graphs and the telegrapher's equation for microwave networks [24,26]. Microwave networks are the only ones that allow for the experimental simulation of quantum systems with all three types of symmetry within the framework of the random matrix theory (RMT): Gaussian orthogonal ensemble (GOE)—systems with preserved time reversal symmetry (TRS) [16,21,24,25,27,30–32], Gaussian unitary ensemble (GUE)—systems with broken TRS [24,28,33–36], and Gaussian symplectic ensemble (GSE)—systems with TRS and half-spin [37]. The other model systems, which are not as versatile as microwave networks, but are often used in simulations of complex quantum systems, are flat microwave billiards [38–54], and exited atoms in strong microwave fields [55–67].

In this article, we will analyze the splitting of a quantum graph (network) into two disconnected subgraphs (subnetworks). Using a currently introduced spectral invariant—the generalized Euler characteristic $\mathcal{E}$ [22]—we determine the number $|V_c|$ of common vertices where the two subgraphs were initially connected. The application of the generalized Euler characteristic $\mathcal{E}$ for this purpose stems from the fact that it can be evaluated without knowing the topologies of quantum graphs (networks), using small or moderate numbers of their lowest eigenenergies (resonances). The theoretical results are numerically verified and confirmed experimentally using the spectra of microwave networks simulating quantum graphs.

2. Theoretical Outline

2.1. The Generalized EULER Characteristic

In Refs. [21,22], the formulas for the Euler characteristic for graphs with the standard boundary conditions at the vertices and with the mixed ones, standard and Dirichlet boundary conditions at vertices, were derived. In the case of the standard boundary conditions,

$$\chi = 2 + 8\pi^2 \sum_{\substack{k_n \in \Sigma(L^{st}(\Gamma)) \\ k_n \neq 0}} \frac{\sin(k_n/t)}{(k_n/t)\left((2\pi)^2 - (k_n/t)^2\right)}|_{t \geq t_0}, \tag{4}$$

where $\Sigma(L^{st}(\Gamma))$ denotes the spectrum of the Laplacian $L^{st}(\Gamma)$ with the standard vertex conditions, taken in the square root scale, i.e., the numbers k_n are the square roots of the eigenenergies λ_n and t is a scaling parameter [19–21] with $t_0 = \frac{1}{2l_{min}}$, where l_{min} is the length of the shortest edge of the graph. The above formula is equivalent to Equation (3); however, instead of using topological information about graphs or networks, such as the number of vertices $|V|$ and edges $|E|$, it requires a certain number of the lowest eigenenergies (resonances) of graphs or networks.

For graphs and networks with the mixed boundary conditions, namely the standard and Dirichlet ones ($|V_D| \neq 0$), the generalized Euler characteristic can be expressed by the following formula:

$$\chi_G := \chi - |V_D| = 8\pi^2 \sum_{k_n \in \Sigma(L^{st,D}(\Gamma))} \frac{\sin(k_n/t)}{(k_n/t)\left((2\pi)^2 - (k_n/t)^2\right)}|_{t \geq t_0}. \tag{5}$$

In Equation (5), the spectrum of the Laplacian $L^{st,D}(\Gamma)$ with the standard and Dirichlet vertex conditions is denoted by $\Sigma(L^{st,D}(\Gamma))$.

The above two equations can be unified into a single one for the generalized Euler characteristic:

$$\mathcal{E}(|V_D|) = 2\delta_{0,|V_D|} + 8\pi^2 \sum_{\substack{k_n \in \Sigma(L(\Gamma)) \\ k_n \neq 0}} \frac{\sin(k_n/t)}{(k_n/t)\left((2\pi)^2 - (k_n/t)^2\right)}|_{t \geq t_0}. \tag{6}$$

Depending on the boundary conditions, $\Sigma(L(\Gamma))$ denotes either the spectrum of the Laplacian $L^{st}(\Gamma)$ or $L^{st,D}(\Gamma)$. In the borderline cases $|V_D| = 0$ and $|V_D| \neq 0$, $\mathcal{E}(|V_D| = 0) = \chi$ and $\mathcal{E}(|V_D| \neq 0) = \chi_G$, recovering, respectively, Equations (4) and (5).

From the experimental point of view, the usefulness of Equation (6) stems from the fact that the generalized Euler characteristic can be evaluated using only a limited number $K = K_{min}$ of the lowest eigenvalues (resonances) [21,22,68,69]

$$K \geq |V| + 2\mathcal{L}t\left[1 - \exp\left(\frac{-\epsilon\pi}{\mathcal{L}t}\right)\right]^{-1/2}, \tag{7}$$

where $|V|$ is the total number of graph vertices, $\mathcal{L} = \sum_{e \in E} l_e$ is the total length of the graph, and ϵ is the accuracy of determining the Euler characteristic from Formula (7). To obtain the smallest possible number of resonances K_{min}, for a given accuracy ϵ, we assign to t its smallest allowed value $t = t_0 = \frac{1}{2l_{min}}$. Since the Euler characteristic is an integer, the accuracy of its determination should be taken $\epsilon < 1/2$. In our calculations of K_{min}, we assumed $\epsilon = 1/4$.

2.2. *A Graph Split into Two Disconnected Subgraphs*

In order to simplify the description of the graphs, we introduce the following notation of graphs and networks $\Gamma(|V|, |E|, |V_D|)$, where $|V| = |V_N| + |V_D|$. A graph or network $\Gamma(|V|, |E|, |V_D|)$ contains $|V|$ vertices, including $|V_N|$ and $|V_D|$ vertices with standard (Neumann) and Dirichlet boundary conditions and $|E|$ edges.

We will consider a general situation when an original graph $\Gamma_o(|V_o|, |E_o|, |V_{D_o}|)$ is split into two disconnected subgraphs $\Gamma_i(|V_i|, |E_i|, |V_{D_i}|)$, $i = 1, 2$, at the common for the subgraphs vertices V_c, which are characterized by the Neumann boundary conditions. In the partition process, each common vertex $v \in V_c$ will be split into two new vertices belonging to the different subgraphs (see Figure 1).

The generalized Euler characteristics of the original graph and its subgraphs are $\mathcal{E}_o(|V_{D_o}|) = |V_o| - |E_o| - |V_{D_o}|$ and $\mathcal{E}_i(|V_{D_i}|) = |V_i| - |E_i| - |V_{D_i}|$, $i = 1, 2$, respectively.

The relationships between the number of vertices and edges of the graphs are the following: $|V_o| + |V_c| = |V_1| + |V_2|$, $|E_o| = |E_1| + |E_2|$. It leads to the following relationship between $\mathcal{E}_o(|V_{D_o}|)$ and $\mathcal{E}_i(|V_{D_i}|)$, $i = 1,2$

$$\mathcal{E}_1(|V_{D_1}|) + \mathcal{E}_2(|V_{D_2}|) = \mathcal{E}_o(|V_{D_o}|) + |V_c| + |V_{D_o}| - |V_{D_1}| - |V_{D_2}|, \tag{8}$$

where $|V_c|$ denotes the number of common vertices.

In Figure 1, we show the case when the original graph $\Gamma_o(|V_o| = 6, |E_o| = 9, |V_{D_o}| = 0) = \Gamma_o(6,9,0)$ is divided into two subgraphs $\Gamma_1(4,6,0)$ and $\Gamma_2(4,3,0)$. Using Equation (8), one can find that the subgraphs before the disconnection were connected in $|V_c| = 2$ common vertices. In this relatively simple situation, the generalized Euler characteristics of the graphs or networks can be found from their topological properties, i.e., the numbers of vertices and edges of the graphs. However, if we do not see the graphs and therefore do not know their topological properties but we know their eigenvalues (spectra), the only available solution to the problem is to use Equation (6) to find their generalized Euler characteristics and consequently the number $|V_c|$ of the common vertices. The same situation exists for the graphs possessing the Dirichlet boundary conditions. In this case, in order to identify them, one needs to know (measure) the eigenvalues (resonances) of graphs or networks and use Equations (6) and (8) to evaluate the number $|V_c|$ of the common vertices.

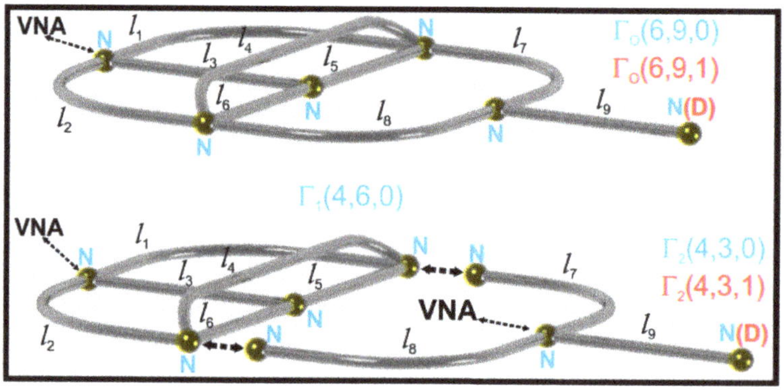

Figure 1. The scheme of the original graph $\Gamma_o(6,9,0)$, which was divided into two subgraphs $\Gamma_1(4,6,0)$ and $\Gamma_2(4,3,0)$. All graphs possess the vertices with the Neumann boundary conditions, which are marked by blue capital letters N. In the case of the graphs with the mixed boundary conditions, the original graph $\Gamma_o(6,9,1)$ was divided into two subgraphs $\Gamma_1(4,6,0)$ and $\Gamma_2(4,3,1)$. The vertices with the Dirichlet boundary conditions are marked by red capital letters D. The vertices where a vector network analyzer was connected to the microwave networks simulating quantum graphs presented in this figure are marked by VNA.

3. Measurements of the Spectra of Microwave Networks

In order to evaluate the generalized Euler characteristic $\mathcal{E}(|V_D|)$ defined by Equation (6), we measured the spectra of microwave networks simulating quantum graphs. In our investigations, we used a set-up (see Figure 2) that consisted of an Agilent E8364B vector network analyzer (VNA) and HP 85133-60016 flexible microwave cable that connected the VNA to the measured network. The flexible cable connected to the network is equivalent to attaching an infinite lead to the quantum graph [22,32]. In this way, the one-port scattering matrix $S_{11}(\nu)$ of the network was measured as a function of microwave frequency ν. The modulus of $|S_{11}(\nu)|$ was used to identify the network's resonances. In Figure 2, we also show the original microwave network $\Gamma_o(6,9,1)$, which possesses a single vertex with the Dirichlet boundary condition ($V_{D_o} = 1$), marked by the red capital letter D. The measured spectrum of the network $\Gamma_o(6,9,1)$ is shown in the inset of Figure 2 in the frequency range $\nu = [0.01, 1]$ GHz. In order to reconfirm our experimental results, the spectra of the quantum graphs simulated by the microwave networks were also calculated numerically using the pseudo-orbits method developed in Ref. [31].

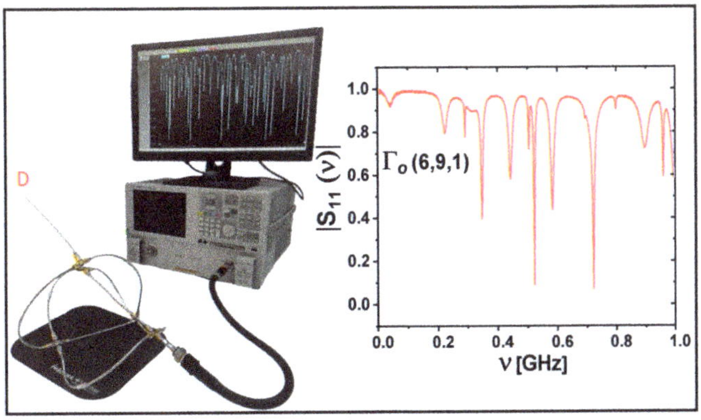

Figure 2. The experimental set-up. It contains an Agilent E8364B vector network analyzer (VNA) and HP 85133-60016 flexible microwave cable that connects the VNA to the measured network. The original microwave network $\Gamma_o(6,9,1)$ possesses a single vertex with the Dirichlet boundary condition, which is marked by the red capital letter D. The measured spectrum of the network $\Gamma_o(6,9,1)$ is shown in the inset in the frequency range $\nu = [0.01, 1]$ GHz.

In the construction of microwave networks simulating quantum graphs, we used microwave coaxial cables and junctions that corresponded to the edges and vertices of the quantum graphs. The microwave cables consisted of an outer conductor with an inner radius $r_2 = 0.15$ cm and an inner conductor of radius $r_1 = 0.05$ cm, which was surrounded by the dielectric material (Teflon). The fundamental TEM mode propagates in such cables below the cut-off frequency of the TE_{11} mode $\nu_{cut} = \frac{c}{\pi(r_1+r_2)\sqrt{\varepsilon}} = 33$ GHz [70,71], where the dielectric constant of Teflon $\varepsilon = 2.06$. It is important to point out that the lengths of edges of the simulated quantum graph have to be compared to the optical lengths of the edges of the microwave networks, i.e., $l_{opt} = \sqrt{\varepsilon} l_{ph}$, where l_{ph} is the physical length of the network edges.

In this paper, we discuss two general situations that are possible when the original network (graph) is split into two subnetworks (subgraphs): the case when the original network and its subnetworks have only the standard boundary conditions and the case when they are characterized by the mixed boundary conditions, when the Dirichlet boundary conditions are present.

3.1. Networks with the Standard Boundary Conditions

Here, we will consider the original network $\Gamma_o(|V_o|, |E_o|, |V_{D_o}|)$, which is split into two disconnected subnetworks $\Gamma_i(|V_i|, |E_i|, |V_{D_i}|)$, $i = 1, 2$, at the common for the subnetworks vertices $v \in V_c$. All networks are characterized by the standard (Neumann) boundary conditions. The experimental realizations of the networks $\Gamma_o(6,9,0)$ and its two subnetworks $\Gamma_1(4,6,0)$ and $\Gamma_2(4,3,0)$ are schematically shown in Figures 1 and 2. In this case, all networks possess only standard (Neumann) boundary conditions, denoted with the capital letter N.

The total optical lengths of the networks $\Gamma_o(6,9,0)$, $\Gamma_1(4,6,0)$, and $\Gamma_2(4,3,0)$ are $\mathcal{L}_o = 2.579$ m, $\mathcal{L}_1 = 1.675$ m, and $\mathcal{L}_2 = 0.940$ m, respectively. The lengths of their shortest edges are $l_{min_o} = l_6 = 0.221$ m, $l_{min_1} = l_6 = 0.221$ m, and $l_{min_2} = l_9 = 0.270$ m, giving $K_{min_o} = 38$, $K_{min_1} = 23$, and $K_{min_2} = 8$, respectively, which were estimated using Equation (7). Experimentally, in order to find the minimum number of resonances determined by the parameters K_{min_o}, K_{min_1}, and K_{min_2}, it was necessary to measure the spectra of the microwave networks $\Gamma_o(6,9,0)$, $\Gamma_1(4,6,0)$, and $\Gamma_2(4,3,0)$ in the frequency ranges

$[0.010, 2.347]$ GHz, $[0.010, 2.234]$ GHz, and $[0.010, 1.271]$ GHz, respectively. Taking into account the above parameters, the generalized Euler characteristics $\mathcal{E}_o(|V_{D_o}|)$, $\mathcal{E}_1(|V_{D_1}|)$, and $\mathcal{E}_2(|V_{D_2}|)$ were calculated using Equation (6).

In Figure 3a–c, we show the generalized Euler characteristics $\mathcal{E}_o(|V_{D_o}| = 0)$, $\mathcal{E}_1(|V_{D_1}| = 0)$, and $\mathcal{E}_2(|V_{D_2}| = 0)$ (red dotted lines), evaluated experimentally as a function of the parameter t. The numerically found generalized Euler characteristics are marked with blue full lines. In all three cases, for both experimental and theoretical results, the plateaus at the generalized Euler characteristics start close to the points $t_{0_o} = 2.26\ \mathrm{m}^{-1}$, $t_{0_1} = 2.26\ \mathrm{m}^{-1}$, and $t_{0_2} = 1.85\ \mathrm{m}^{-1}$ defined by the theory (see the discussion below Equation (7)). The values of the generalized Euler characteristics are found to be $\mathcal{E}_o(|V_{D_o}| = 0) = -3$, $\mathcal{E}_1(|V_{D_1}| = 0) = -2$, and $\mathcal{E}_2(|V_{D_2}| = 0) = 1$, respectively. Using Equation (8), it is easy to find that $|V_c| = 2$. It means that, before splitting, the two subgraphs were connected at the two vertices. It is important to point out that the above information was obtained without knowing anything about the topologies of the networks.

3.2. Networks with the Mixed Boundary Conditions

We used the same physical networks to investigate the split of the original network $\Gamma_o(6, 9, 1)$ possessing the mixed boundary conditions into two separated subnetworks $\Gamma_1(4, 6, 0)$ and $\Gamma_2(4, 3, 1)$. The network $\Gamma_o(6, 9, 1)$ and the subnetwork $\Gamma_2(4, 3, 1)$ possess a single Dirichlet boundary condition. Figure 1 shows the schemes of the networks. The Dirichlet boundary conditions are denoted by the capital letter D. All other parameters of the networks, such as the total lengths and the shortest edges, are the same as in the case of the networks with the standard boundary conditions, which were discussed above. However, for the networks with the mixed boundary conditions, one requires the same number of resonances as, in the case of the networks with the Neumann boundary conditions, the frequency ranges where they can be identified are different. For example, for the networks $\Gamma_o(6, 9, 1)$ and $\Gamma_2(4, 3, 1)$, they are $[0.010, 2.500]$ GHz and $[0.010, 1.131]$ GHz, respectively.

In Figure 4a–c, we show the generalized Euler characteristics $\mathcal{E}_o(|V_{D_o}| = 1)$, $\mathcal{E}_1(|V_{D_1}| = 0)$, and $\mathcal{E}_2(|V_{D_2}| = 1)$ (red dotted lines), evaluated experimentally as a function of the parameter t. The generalized Euler characteristics that were found numerically are marked with blue full lines. Moreover, here, in all three cases, for both experimental and theoretical results, the plateaus at the generalized Euler characteristics start close to the points t_{0_o}, t_{0_1}, and t_{0_2} defined by the theory. The values of the generalized Euler characteristics are found to be $\mathcal{E}_o(|V_{D_o}| = 1) = -4$, $\mathcal{E}_1(|V_{D_1}| = 0) = -2$, and $\mathcal{E}_2(|V_{D_2}| = 1) = 0$, respectively. In addition, in this case, using Equation (8), we found that $|V_c| = 2$. One should remark that in the case of the mixed boundary conditions, the knowledge of the topologies of the experimental networks does not allow us to find their generalized Euler characteristics. We also have to know the number of their Dirichlet boundary conditions. Therefore, the measurements of the spectra of the networks and using Equation (6) are mandatory.

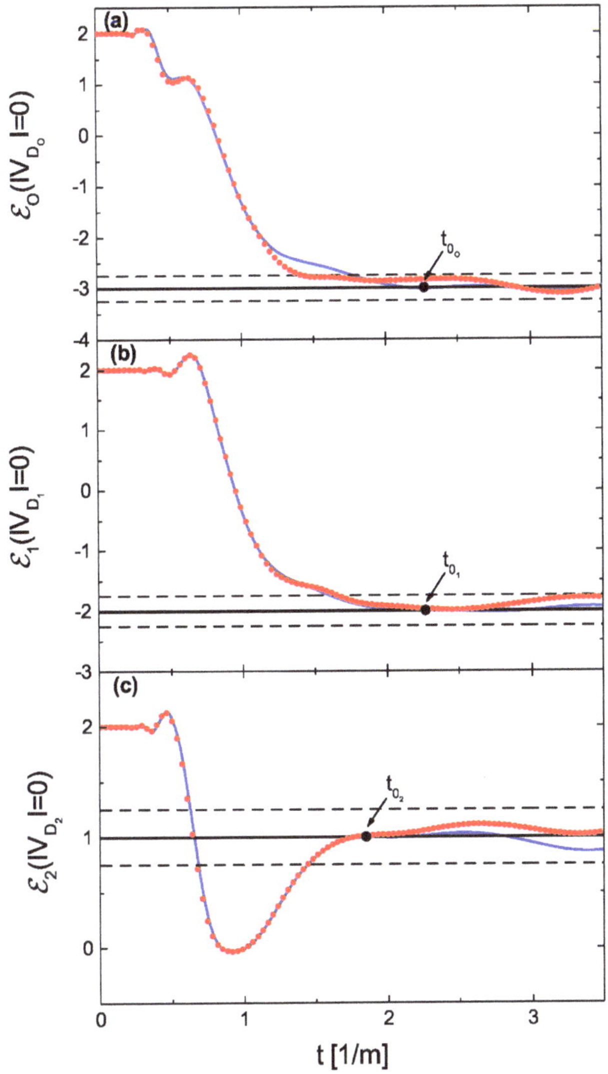

Figure 3. Generalized Euler characteristics evaluated for the networks with the standard boundary conditions as a function of the parameter t. Panels (**a–c**) show the generalized Euler characteristics $\mathcal{E}_o(|V_{D_o}|=0)$, $\mathcal{E}_1(|V_{D_1}|=0)$, and $\mathcal{E}_2(|V_{D_2}|=0)$ of the networks $\Gamma_o(6,9,0)$, $\Gamma_1(4,6,0)$, and $\Gamma_2(4,3,0)$, respectively. The experimental and numerical results are marked with red dotted and blue full lines, respectively. In all three cases, the plateaus at the generalized Euler characteristics start close to the points $t_{0_o} = 2.26\ \mathrm{m}^{-1}$, $t_{0_1} = 2.26\ \mathrm{m}^{-1}$, and $t_{0_2} = 1.85\ \mathrm{m}^{-1}$, respectively, defined by the theory (see the discussion below Equation (7)). The black broken lines show the limits of the expected errors $\mathcal{E}_q(|V_{D_q}|) \pm 1/4$, where $q = o, 1$, and 2.

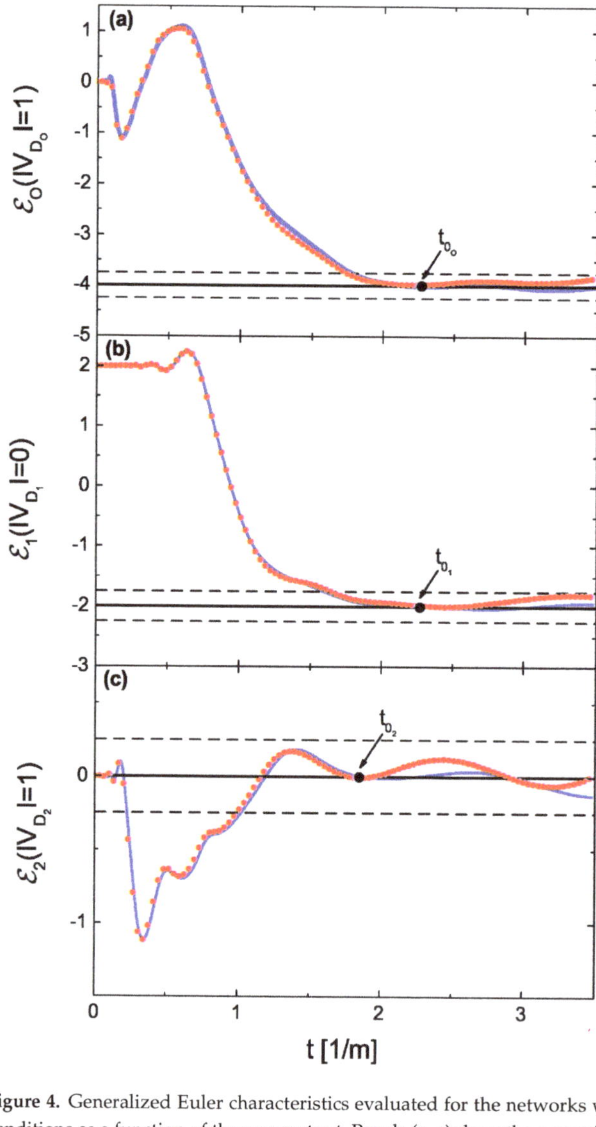

Figure 4. Generalized Euler characteristics evaluated for the networks with the mixed boundary conditions as a function of the parameter t. Panels (**a–c**) show the generalized Euler characteristics $\mathcal{E}_o(|V_{D_o}| = 1)$, $\mathcal{E}_1(|V_{D_1}| = 0)$, and $\mathcal{E}_2(|V_{D_2}| = 1)$ of the networks $\Gamma_o(6,9,1)$, $\Gamma_1(4,6,0)$, and $\Gamma_2(4,3,1)$, respectively. The experimental and numerical results are marked with red dotted and blue full lines, respectively. Moreover, here, in all three cases, the plateaus at the generalized Euler characteristics start close to the points $t_{0_o} = 2.26\ \mathrm{m}^{-1}$, $t_{0_1} = 2.26\ \mathrm{m}^{-1}$, and $t_{0_2} = 1.85\ \mathrm{m}^{-1}$, respectively, defined by the theory. The black broken lines show the limits of the expected errors $\mathcal{E}_q(|V_{D_q}|) \pm 1/4$, where $q = o, 1$, and 2.

4. Summary

We analyzed a relationship between the generalized Euler characteristic $\mathcal{E}_o(|V_{D_o}|)$ of the original graph (network), which was split into two disconnected subgraphs (subnetworks) $i = 1, 2$, and their generalized Euler characteristics $\mathcal{E}_i(|V_{D_i}|)$. We showed that the evaluation of the generalized Euler characteristics $\mathcal{E}_o(|V_{D_o}|)$ and $\mathcal{E}_i(|V_{D_i}|)$ allows us to determine the number $|V_c|$ of common vertices where the two subgraphs were initially connected. The theoretical results were numerically verified and confirmed experimentally using microwave networks with the standard and mixed boundary conditions. The application of the generalized Euler characteristics defined by Equation (6) requires the measurement of the spectra of the networks but in return allows us to find $|V_c|$ without knowing their topologies. Therefore, it might be possible to apply the properties of the splitting networks discussed in this article in some more practical applications, such as the diagnostics of electronic or microwave networks. One should underline that the first practical test of such diagnostics where the properties of splitting networks and the generalized Euler characteristic were applied was presented in this article. For this purpose, we used real-world systems, such as microwave networks. They are open and dissipative systems, which are completely different from the ideal dissipationless graphs considered in their mathematical studies. In spite of this, even for more complex networks possessing the mixed boundary conditions, we were able to find experimentally the number of common vertices $|V_c|$ where the two separated subnetworks were connected before their splitting.

Author Contributions: O.F. performed the experiment. M.Ł. performed numerical calculations. O.F., M.Ł., A.A. and S.B. performed the data analysis. L.S. provided the theoretical interpretation. L.S. wrote the manuscript. All authors have read and agreed to the published version of the manuscript.

Funding: This research study was funded by the National Science Centre, Poland, Grant No. 2018/30/Q/ ST2/00324.

Institutional Review Board Statement: Not applicable.

Informed Consent Statement: Not applicable.

Data Availability Statement: The data that support the results presented in this paper and other findings of this study are available from the corresponding authors upon reasonable request.

Conflicts of Interest: The authors declare no conflict of interest.

References

1. Euler, L. Solutio problematis ad geometriam situs pertinentis. *Comment. Acad. Sci. Imp. Petropol.* **1736**, *8*, 128–140.
2. Pauling, L. The diamagnetic anisotropy of aromatic molecules. *J. Chem. Phys.* **1936**, *4*, 673–677. [CrossRef]
3. Sánchez-Gil, J.A.; Freilikher, V.; Yurkevich, I.; Maradudin, A.A. Coexistence of ballistic transport, diffusion, and localization in surface disordered waveguides. *Phys. Rev. Lett.* **1998**, *80*, 948–953. [CrossRef]
4. Kowal, D.; Sivan, U.; Entin-Wohlman, O.; Imry, Y. Transmission through multiply-connected wire systems. *Phys. Rev. B* **1990**, *42*, 9009–9018. [CrossRef] [PubMed]
5. Imry, Y. *Introduction to Mesoscopic Physics*; Oxford University Press: New York, NY, USA, 1997; p. 234.
6. Hu, W.; Pillay, J.C.; Wu, K.; Pasek, M.; Shum, P.P.; Chong, Y.D. Measurement of a topological edge invariant in a microwave network. *Phys. Rev. X* **2015**, *5*, 011012. [CrossRef]
7. Szameit, A. Photonics: Chaos from symmetry. *Nat. Phys.* **2015**, *11*, 895–896. [CrossRef]
8. Exner, P.; Šeba, P.; Stovicek, P. Quantum interference on graphs controlled by an external electric field. *J. Phys. A Math. Gen.* **1988**, *21*, 4009–4019. [CrossRef]
9. Kottos, T.; Smilansky, U. Quantum chaos on graphs. *Phys. Rev. Lett.* **1997**, *79*, 4794–4797. [CrossRef]
10. Kottos, T.; Smilansky, U. Periodic Orbit Theory and Spectral Statistics for Quantum Graphs. *Ann. Phys.* **1999**, *274*, 76–124. [CrossRef]
11. Blümel, R.; Dabaghian, Y.; Jensen, R.V. Explicitly Solvable Cases of One-Dimensional Quantum Chaos. *Phys. Rev. Lett.* **2002**, *88*, 4. [CrossRef]
12. Berkolaiko, G.; Kuchment, P. Introduction to Quantum Graphs. In *Mathematical Surveys and Monographs;* American Mathematical Society, Providence, RI, USA, 2013; p. 270.
13. Pluhar, Z.; Weidenmueller, H.A. Universal Quantum Graphs. *Phys. Rev. Lett.* **2014**, *112*, 144102. [CrossRef] [PubMed]
14. Pinheiro, L.K.; Souza, B.S.; Trevisan, V. Determining Graphs by the Complementary Spectrum. *Discuss. Math.—Graph Theory* **2020**, *40*, 607–620. [CrossRef]
15. Białous, M.; Dulian, P.; Sawicki, A.; Sirko, L. Delay-time distribution in the scattering of short Gaussian pulses in microwave networks. *Phys. Rev. E* **2021**, *104*, 024223. [CrossRef] [PubMed]
16. Hul, O.; Ławniczak, M.; Bauch, S.; Sawicki, A.; Kuś, M.; Sirko, L. Are scattering properties of graphs uniquely connected to their shapes? *Phys. Rev. Lett.* **2012**, *109*, 040402. [CrossRef]

17. Gadomski, A.; Łuczka, J.; Rudnicki, R. Finite volume effects in a model grain growth. *Phys. A Stat. Mech. Appl.* **2003**, *325*, 284–291. [CrossRef]
18. Gadomski, A.; Kruszewska, N. On clean grain-boundaries involving growth of nonequilibrium crystalline-amorphous superconducting materials addressed by a phenomenological viewpoint. *Eur. Phys. J. B* **2012**, *85*, 1–8. [CrossRef]
19. Kurasov, P. Graph Laplacians and topology. *Ark. Mat.* **2008**, *46*, 95–111. [CrossRef]
20. Kurasov, P. Schrödinger operators on graphs and geometry I: Essentially bounded potentials. *J. Funct. Anal.* **2008**, *254*, 934–953. [CrossRef]
21. Ławniczak, M.; Kurasov, P.; Bauch, S.; Białous, M.; Yunko, V.; Sirko, L. Hearing Euler characteristic of graphs. *Phys. Rev. E* **2020**, *101*, 052320. [CrossRef]
22. Ławniczak, M.; Kurasov, P.; Bauch, S.; Białous, M.; Akhshani, A.; Sirko, L. A new spectral invariant for quantum graphs. *Sci. Rep.* **2021**, *11*, 15342. [CrossRef]
23. Bauch, S.; Ławniczak, M.; Wrochna, J.; Kurasov, P.; Sirko, L. Some Applications of Generalized Euler Characteristic of Quantum Graphs and Microwave Networks. *Acta Phys. Pol. A* **2021**, *140*, 525–531. [CrossRef]
24. Hul, O.; Bauch, S.; Pakoński, P.; Savytskyy, N.; Życzkowski, K.; Sirko, L. Experimental simulation of quantum graphs by microwave networks. *Phys. Rev. E* **2004**, *69*, 7. [CrossRef] [PubMed]
25. Ławniczak, M.; Hul, O.; Bauch, S.; Šeba, P.; Sirko, L. Experimental and numerical investigation of the reflection coefficient and the distributions of Wigner's reaction matrix for irregular graphs with absorption. *Phys. Rev. E* **2008**, *77*, 056210. [CrossRef] [PubMed]
26. Ławniczak, M.; Bauch, S.; Sirko, L. Application of Microwave Networks to Simulation of Quantum Graphs. In *Handbook of Applications of Chaos Theory*; Chapman and Hall/CRC: New York, NY, USA, 2016; pp. 559–584.
27. Dietz, B.; Yunko, V.; Białous, M.; Bauch, S.; Ławniczak, M.; Sirko, L. Nonuniversality in the spectral properties of time-reversal-invariant microwave networks and quantum graphs. *Phys. Rev. E* **2017**, *95*, 052202. [CrossRef] [PubMed]
28. Ławniczak, M.; Sirko, L. Investigation of the diagonal elements of the Wigner's reaction matrix for networks with violated time reversal invariance. *Sci. Rep.* **2019**, *9*, 5630. [CrossRef] [PubMed]
29. Yunko, V.; Białous, M.; Sirko, L. Edge switch transformation in microwave networks. *Phys. Rev. E* **2020**, *102*, 012210. [CrossRef]
30. Hul, O.; Tymoshchuk, O.; Bauch, S.; Koch, P.M.; Sirko, L. Experimental investigation of Wigner's reaction matrix for irregular graphs with absorption. *J. Phys. A Math. Theor.* **2005**, *38*, 10489. [CrossRef]
31. Lipovský, J. On the effective size of a non-Weyl graph. *J. Phys. A Math. Theor* **2016**, *49*, 375202. [CrossRef]
32. Ławniczak, M.; Lipovský, J.; Sirko, L. Non-Weyl microwave graphs. *Phys. Rev. Lett.* **2019**, *122*, 140503. [CrossRef]
33. Ławniczak, M.; Bauch, S.; Hul, O.; Sirko, L. Experimental investigation of the enhancement factor for microwave irregular networks with preserved and broken time reversal symmetry in the presence of absorption. *Phys. Rev. E* **2010**, *81*, 046204. [CrossRef]
34. Allgaier, M.; Gehler, S.; Barkhofen, S.; Stöckmann, H.J.; Kuhl, U. Spectral properties of microwave graphs with local absorption. *Phys. Rev. E* **2014**, *89*, 022925. [CrossRef] [PubMed]
35. Białous, M.; Yunko, V.; Bauch, S.; Ławniczak, M.; Dietz, B.; Sirko, L. Power Spectrum Analysis and Missing Level Statistics of Microwave Graphs with Violated Time Reversal Invariance. *Phys. Rev. Lett.* **2016**, *117*, 144101. [CrossRef] [PubMed]
36. Ławniczak, M.; Białous, M.; Yunko, V.; Bauch, S.; Dietz, B.; Sirko, L. Analysis of missing level statistics for microwave networks simulating quantum chaotic graphs without time reversal symmetry—The case of randomly lost resonances. *Acta Phys. Pol. A* **2017**, *132*, 1672–1676. [CrossRef]
37. Rehemanjiang, A.; Allgaier, M.; Joyner, C.H.; Müller, S.; Sieber, M.; Kuhl, U.; Stöckmann, H.J. Microwave Realization of the Gaussian Symplectic Ensemble. *Phys. Rev. Lett.* **2016**, *117*, 064101. [CrossRef] [PubMed]
38. Dietz, B.; Friedrich, T.; Harney, H.L.; Miski-Oglu, M.; Richter, A.; Schäfer, F.; Weidenmüller, H.A. Quantum chaotic scattering in microwave resonators. *Phys. Rev. E* **2010**, *81*, 036205. [CrossRef]
39. Yeh, J.H.; Drikas, Z.; Gil Gil, J.; Hong, S.; Taddese, B.; Ott, E.; Antonsen, T.; Andreadis, T.; Anlage, S. Impedance and Scattering Variance Ratios of Complicated Wave Scattering Systems in the Low Loss Regime. *Acta Phys. Pol. A* **2013**, *124*, 1045–1052. [CrossRef]
40. Zheng, X.; Hemmady, S.; Antonsen, T.M.; Anlage, S.M.; Ott, E. Characterization of fluctuations of impedance and scattering matrices in wave chaotic scattering. *Phys. Rev. E* **2006**, *73*, 046208. [CrossRef]
41. Stöckmann, H.J.; Stein, J. "Quantum" chaos in billiards studied by microwave absorption. *Phys. Rev. Lett.* **1990**, *64*, 2215–2218. [CrossRef]
42. Sridhar, S.; Kudrolli, A. Experiments on not "hearing the shape" of drums. *Phys. Rev. Lett.* **1994**, *72*, 2175–2178. [CrossRef]
43. Sirko, L.; Koch, P.M.; Blümel, R. Experimental Identification of Non-Newtonian Orbits Produced by Ray Splitting in a Dielectric-Loaded Microwave Cavity. *Phys. Rev. Lett.* **1997**, *78*, 2940–2943. [CrossRef]
44. Hlushchuk, Y.; Kohler, A.; Bauch, S.; Sirko, L.; Blümel, R.; Barth, M.; Stöckmann, H.J. Autocorrelation function of level velocities for ray-splitting billiards. *Phys. Rev. E* **2000**, *61*, 366–370. [CrossRef] [PubMed]
45. Hlushchuk, Y.; Błedowski, A.; Savytskyy, N.; Sirko, L. Numerical Investigation of Regimes of Wigner and Shnirelman Ergodicity in Rough Billiards. *Phys. Scr.* **2001**, *64*, 192–196. [CrossRef]
46. Hlushchuk, Y.; Sirko, L.; Kuhl, U.; Barth, M.; Stöckmann, H.J. Experimental investigation of a regime of Wigner ergodicity in microwave rough billiards. *Phys. Rev. E* **2001**, *63*, 046208. [CrossRef]

47. Blümel, R.; Koch, P.M.; Sirko, L. Ray-Splitting Billiards. *Foun. Phys.* **2001**, *31*, 269–281. [CrossRef]
48. Dhar, A.; Madhusudhana Rao, D.; Shankar, U.; Sridhar, S. Isospectrality in chaotic billiards. *Phys. Rev. E* **2003**, *68*, 5. [CrossRef] [PubMed]
49. Savytskyy, N.; Hul, O.; Sirko, L. Experimental investigation of nodal domains in the chaotic microwave rough billiard. *Phys. Rev. E* **2004**, *70*, 6. [CrossRef]
50. Hemmady, S.; Zheng, X.; Ott, E.; Antonsen, T.M.; Anlage, S.M. Universal impedance fluctuations in wave chaotic systems. *Phys. Rev. Lett.* **2005**, *94*, 014102. [CrossRef]
51. Hul, O.; Savytskyy, N.; Tymoshchuk, O.; Bauch, S.; Sirko, L. Investigation of nodal domains in the chaotic microwave ray-splitting rough billiard. *Phys. Rev. E* **2005**, *72*, 066212. [CrossRef]
52. Dietz, B.; Richter, A. Quantum and wave dynamical chaos in superconducting microwave billiards. *Chaos* **2015**, *25*, 97601. [CrossRef]
53. Białous, M.; Dietz, B.; Sirko, L. Experimental investigation of the elastic enhancement factor in a microwave cavity emulating a chaotic scattering system with varying openness. *Phys. Rev. E* **2019**, *100*, 012210. [CrossRef]
54. Dietz, B.; Klaus, T.; Miski-Oglu, M.; Richter, A.; Wunderle, M. Partial time-reversal invariance violation in a flat, superconducting microwave cavity with the shape of a chaotic Africa billiard. *Phys. Rev. Lett.* **2019**, *123*, 174101. [CrossRef] [PubMed]
55. Blümel, R.; Buchleitner, A.; Graham, R.; Sirko, L.; Smilansky, U.; Walther, H. Dynamical localization in the microwave interaction of Rydberg atoms: The influence of noise. *Phys. Rev. A* **1991**, *44*, 4521. [CrossRef] [PubMed]
56. Jensen, R.V.; Susskind, S.M.; Sanders, M.M. Chaotic ionization of highly excited hydrogen atoms: Comparison of classical and quantum theory with experiment. *Phys. Rep.* **1991**, *201*, 1–56. [CrossRef]
57. Bellermann, M.; Bergeman, T.; Haffmans, A.; Koch, P.M.; Sirko, L. Electric-field dependence of E1 transitions between highly excited hydrogen Stark sublevels. *Phys. Rev. A* **1992**, *46*, 5836. [CrossRef] [PubMed]
58. Sirko, L.; Yoakum, S.; Haffmans, A.; Koch, P.M. Microwave-driven He Rydberg atoms: Floquet-state degeneracy lifted by a second frequency, Stueckelberg oscillations, and their destruction by added noise. *Phys. Rev. A* **1993**, *47*, R782. [CrossRef]
59. Buchleitner, A.; Delande, D. Quantum dynamics of a circular Rydberg state in a microwave field. *Phys. Rev. Lett.* **1993**, *71*, 3633. [CrossRef]
60. Sirko, L.; Bellermann, M.R.; Haffmans, A.; Koch, P.M.; Richards, D. Probing quantal dynamics of mixed phase space systems with noise. *Phys. Rev. Lett.* **1993**, *71*, 2895. [CrossRef]
61. Bayfield, J.E.; Luie, S.Y.; Perotti, L.C.; Skrzypkowski, M.P. Excited hydrogen atoms in pulsed microwaves: Journeys to quantum chaos and back. *Physics D* **1995**, *83*, 46–54. [CrossRef]
62. Sirko, L.; Koch, P. The pendulum approximation for the main quantal resonance zone in periodically driven hydrogen atoms. *Appl. Phys. B* **1995**, *60*, S195.
63. Sirko, L.; Haffmans, A.; Bellermann, M.R.; Koch, P.M. Microwave "ionization" of excited hydrogen atoms: Frequency dependence in a resonance zone. *Eur. Lett.* **1996**, *33*, 181. [CrossRef]
64. Kaulakys, B.; Ciziunas, A. A theoretical determination of the diffusion-like ionisation time of Rydberg atoms. *J. Phys. B Atom. Mol. Phys.* **1987**, *20*, 1031. [CrossRef]
65. Sirko, L.; Zelazny, S.A.; Koch, P.M. Use of the relative phase in a bichromatic field pulse to control a quasienergy gap. *Phys. Rev. Lett.* **2001**, *87*, 43002-1–43002-4. [CrossRef] [PubMed]
66. Sirko, L.; Koch, P.M. Control of common resonances in bichromatically driven hydrogen atoms. *Phys. Rev. Lett.* **2002**, *89*, 274101. [CrossRef] [PubMed]
67. Arakelyan, A.; Nunkaew, J.; Gallagher, T.F. Ionization of Na Rydberg atoms by a 79-GHz microwave field. *Phys. Rev. A* **2016**, *94*, 053416. [CrossRef]
68. Ławniczak, M.; Kurasov, P.; Bauch, S.; Białous, M.; Sirko, L. The Relationship Between the Euler Characteristic and the Spectra of Graphs and Networks. In Proceedings of the Chaotic Modeling and Simulation International Conference, Florence, Italy, 9–12 June 2020; pp. 487–497.
69. Ławniczak, M.; Kurasov, P.; Bauch, S.; Białous, M.; Sirko, L. Euler characteristic of graphs and networks. *Acta Phys. Pol. A* **2021**, *139*, 323–327. [CrossRef]
70. Jones, D.S. *The Theory of Electromagnetism*; Pergamon Press: Oxford, UK, 1964.
71. Savytskyy, N.; Kohler, A.; Bauch, S.; Blümel, R.; Sirko, L. Parametric correlations of the energy levels of ray-splitting billiards. *Phys. Rev. E* **2001**, *64*, 5. [CrossRef]

MDPI
St. Alban-Anlage 66
4052 Basel
Switzerland
Tel. +41 61 683 77 34
Fax +41 61 302 89 18
www.mdpi.com

Entropy Editorial Office
E-mail: entropy@mdpi.com
www.mdpi.com/journal/entropy

www.ingramcontent.com/pod-product-compliance
Lightning Source LLC
Chambersburg PA
CBHW061615210926
51850CB00009B/2330
* 9 7 8 3 0 3 6 5 5 2 7 5 0 *